T0075241

Crime by the Numbers

Crime by the Numbers: A Criminologist's Guide to R introduces the programming language R and covers the necessary skills to conduct quantitative research in criminology. By the end of this book, a person without any prior programming experience can take raw crime data, be able to clean it, visualize the data, present it using R Markdown, and change it to a format ready for analysis. *Crime by the Numbers* focuses on skills specifically for criminology such as spatial joins, mapping, and scraping data from PDFs, however any social scientist looking for an introduction to R for data analysis will find this useful.

Key Features:

- Introduction to RStudio including how to change user preference settings
- Basic data exploration and cleaning – subsetting, loading data, regular expressions, aggregating data
- Graphing with ggplot2
- How to make maps (hotspot maps, choropleth maps, interactive maps)
- Webscraping and PDF scraping
- Project management – how to prepare for a project, how to decide which projects to do, best ways to collaborate with people, how to store your code (using git), and how to test your code.

Jacob Kaplan is the Chief Data Scientist of the Research on Policing Reform and Accountability (RoPRA), a multi-disciplinary, multi-institutional team of social scientists studying the feasibility and efficacy of policing reform, with a focus on statistically rigorous research and practical applications. His current appointment is at the Princeton School of Public and International Affairs. He holds a PhD and a master's degree in criminology from the University of Pennsylvania and a bachelor's degree in criminal justice from California State University, Sacramento. He is the author of several R packages that make it easier to work with data, including fastDummies and asciiSetupReader. He is also the author of books on the two primary criminal justice data sets: the FBI's Uniform Crime Reporting (UCR) Program Data, and the FBI's National Incident Based Reporting System (NIBRS) data.

Chapman & Hall/CRC
The R Series

Series Editors
John M. Chambers, Department of Statistics, Stanford University, California, USA
Torsten Hothorn, Division of Biostatistics, University of Zurich, Switzerland
Duncan Temple Lang, Department of Statistics, University of California, Davis, USA
Hadley Wickham, RStudio, Boston, Massachusetts, USA

Recently Published Titles

Engineering Production-Grade Shiny Apps
Colin Fay, Sébastien Rochette, Vincent Guyader, and Cervan Girard

Javascript for R
John Coene

Advanced R Solutions
Malte Grosser, Henning Bumann, and Hadley Wickham

Event History Analysis with R, Second Edition
Göran Broström

Behavior Analysis with Machine Learning Using R
Enrique Garcia Ceja

Rasch Measurement Theory Analysis in R: Illustrations and Practical Guidance for Researchers and Practitioners
Stefanie Wind and Cheng Hua

Spatial Sampling with R
Dick R. Brus

Crime by the Numbers: A Criminologist's Guide to R
Jacob Kaplan

Analyzing US Census Data: Methods, Maps, and Models in R
Kyle Walker

ANOVA and Mixed Models: A Short Introduction Using R
Lukas Meier

Tidy Finance with R
Stefan Voigt, Patrick Weiss and Christoph Scheuch

Deep Learning and Scientific Computing with R torch
Sigrid Keydana

Model-Based Clustering, Classification, and Density Estimation Using mclust in R
Lucca Scrucca, Chris Fraley, T. Brendan Murphy, and Adrian E. Raftery

For more information about this series, please visit: https://www.crcpress.com/Chapman--HallCRC-The-R-Series/book-series/CRCTHERSER

Crime by the Numbers

A Criminologist's Guide to R

Jacob Kaplan

CRC Press
Taylor & Francis Group
Boca Raton London New York

CRC Press is an imprint of the
Taylor & Francis Group, an **informa** business

A CHAPMAN & HALL BOOK

First edition published 2023
by CRC Press
6000 Broken Sound Parkway NW, Suite 300, Boca Raton, FL 33487-2742

and by CRC Press
4 Park Square, Milton Park, Abingdon, Oxon, OX14 4RN

CRC Press is an imprint of Taylor & Francis Group, LLC

Library of Congress Cataloging-in-Publication Data

Names: Kaplan, Jacob (Criminologist), author.
Title: Crime by the numbers : a criminologist's guide to R : Jacob Kaplan.
Description: First edition. | Boca Raton : CRC Press, [2023] | Series: Chapman & Hall/CRC the R series | Includes bibliographical references and index.
Identifiers: LCCN 2022006254 (print) | LCCN 2022006255 (ebook) | ISBN 9781032244075 (hardback) | ISBN 9781032245515 (paperback) | ISBN 9781003279211 (ebook)
Subjects: LCSH: Criminal statistics--Data processing. | Criminology--Data processing. | R (Computer program language)
Classification: LCC HV7415 .K37 2023 (print) | LCC HV7415 (ebook) | DDC 364.0285--dc23/eng/20220528
LC record available at https://lccn.loc.gov/2022006254
LC ebook record available at https://lccn.loc.gov/2022006255

ISBN: 978-1-032-24407-5 (hbk)
ISBN: 978-1-032-24551-5 (pbk)
ISBN: 978-1-003-27921-1 (ebk)

DOI: 10.1201/9781003279211

Typeset in Latin Modern font
by KnowledgeWorks Global Ltd.

Publisher's note: This book has been prepared from camera-ready copy provided by the authors.

To my love Kristina, and our puppies Peanut and Moose.

Contents

Preface

This book introduces the programming language R and is meant for undergrads or graduate students studying criminology. R is a programming language that is well-suited to the type of work frequently done in criminology - taking messy data and turning it into useful information. While R is a useful tool for many fields of study, this book focuses on the skills criminologists should know and uses crime data for the example data sets.

For this book you should have the latest version of R[1] installed and be running it through RStudio Desktop (the free version).[2] We'll get into detail on what R and RStudio are soon, but please have them both installed to be able to follow along with each chapter. While you must install both, you only ever need to open RStudio. While R is the actual programming language, RStudio is a program that makes it a lot easier to interact with R than opening up the R application itself.[3] I highly recommend following along with the code for each lesson and then trying to use the lessons learned on a data set that you are interested in.

Why learn to program?

With the exception of some more advanced techniques like scraping data from websites or from PDFs, nearly everything we do here can be done through Excel, a software you're probably more familiar with. The basic steps for research projects are generally:

1. Open up a data set - which frequently comes as an Excel file!
2. Change some values - misspellings or too-specific categories for our purposes are very common in crime data
3. Delete some values - such as states you won't be studying
4. Make some graphs
5. Calculate some values - such as number of crimes per year

---

[1] https://cloud.r-project.org/
[2] https://www.rstudio.com/products/rstudio/download/
[3] This is formally known as an "integrated development environment" or an IDE.

6. Sometimes do a statistical analysis depending on the type of project
7. Write up what you find

R can do all of this but why should you want (or have) to learn an entirely new skill just to do something you can already do? R is useful for two main reasons: scale and reproducibility.

Scale

If you do a one-off project in your career such as downloading some data and making a graph out of it, it makes sense to stick with software like Excel. The cost (in time and effort) of learning R is certainly not worth it for a single (or even several) project - even one perfectly suited for using R. R (and many programming languages more generally, such as Python) has its strength in doing something fairly simple many times. For example, it may be quicker to download one file yourself than it is to write the code in R to download that file. But when it comes to downloading hundreds of files, writing the R code becomes very quickly the better option than doing it by hand.

For most tasks you do in research when dealing with data, you will end up doing them many times (including doing the same task in future projects). So R offers the trade-off of spending time upfront by learning the code with the benefit of that code being able to do work at a large scale with little extra work from you. Please keep in mind this trade-off - you need to front-load the costs of learning R for the rewards of making your life easier when dealing with data - when feeling discouraged about the small returns you get early in learning R.

Reproducibility

The second major benefit of using R over something like Excel is that R is reproducible. Every action you take is written down. This is useful when collaborating with others (including your future self) as they can look at your code and follow along what you did without you having to show them every click you made as you frequently would on Excel. Your collaborator can look at your code to help you figure out a bug in the code or add their own code to yours.

In the research context specifically, you want to have code to give to people to ensure that your research was done correctly and there aren't bugs in the code. Additionally, if you build a tool to, for example, interpret raw crime data from an agency and turn it into a map, being able to share the code so others can modify it for their own city saves these people a lot of time and effort.

While not required (yet) in criminology, some academic journals (such as in economics) even require that you submit your data and code if your paper is accepted. If criminology follows in this trend, or if you submit to journals that require code submissions, you'll need to be able to write code and not rely on software that doesn't track your steps (such as Excel and SPSS).

What you will learn

For many of the lessons we will be working through real research questions and working from start to finish as you would on your own project. This involves thinking about what you want to accomplish from the data you have and what steps you need to take to reach that goal. This involves more than just knowing what code to write - it includes figuring out what your data has, whether it can answer the question you're asking, and planning out (without writing any code yet) what you need to do when you start coding. For most lessons we'll be using actual crime data that is commonly used in research so you'll become acquainted with a number of important data sets.

Skills

There is a large range of skills in criminology research - far too large to cover in a single book. Here we will attempt to teach fundamental skills to build a solid foundation for future work. We'll be focusing on the following skills and trying to reinforce our skills with each lesson.

• Subsetting - Taking only certain rows or columns from a data set
• Graphing
• Regular expressions - Essentially R's "Find and Replace" function for text
• Getting data from websites (webscraping)
• Getting data from PDFs (PDF scraping)
• Mapping
• Writing documents through R

What you won't learn

This book is not a statistics book so we will not be covering any statistical techniques. Though some data sets we handle are fairly large, this book does not discuss how to deal with Big Data. While the lessons you learn in this book can apply to larger data sets, Big Data (which I tend to define loosely as data that are too large for my computer to handle) requires special skills that are outside the realm of this book. If you do intend to deal with huge data sets I recommend you look at the R package data.table,[4] which is an excellent resource for it. While we briefly cover mapping, this book will not cover working with geographic data in detail. For a comprehensive look at geographic data please see this book.[5] This book also will not cover any qualitative data or analysis. While qualitative research is an important part of criminology, this book only focuses on working with quantitative data. Some parts of this book may apply to dealing with qualitative data, such as PDF scraping and regular expressions, but the examples I use in those chapters still deal with quantitative data.

Simple vs easy

In the course of this book we will cover things that are very simple. For example, we'll take a data set (think of it like an Excel file) with crime for nearly every police agencyg in the United States and keep only data from Colorado for a small number of years. We'll then find out how many murders happened in Colorado each year. This is a fairly simple task - it can be expressed in two sentences. You'll find that most of what you do is simple like this - it is quick to talk about what you are doing and the concepts are not complicated. What it isn't is easy. To actually write the R code to do this takes knowing a number of interrelated concepts in R and several lines of code to implement each step.

While this distinction may seem minor, I think it is important for newer programmers to understand that what they are doing may be simple to talk about but hard to implement. When you learn something new in R, or are first introduced to the language, you may feel like you're bashing your head through a brick wall. That is normal. It is easy to feel like a bad programmer because something that can be articulated in 10 seconds may take hours to do. So during times when you are working with R try to keep in mind that even though a project may be simple to articulate, it may be hard to code and that there is often very little correlation between the two.

---

[4]https://github.com/Rdatatable/data.table/wiki
[5]https://geocompr.robinlovelace.net/

How to read this book

This book is written so a person who has no programming experience can start with this chapter and by the end of the book be able to do a data project from start to finish. Each chapter introduces a new skill and builds on the skills introduced in previous chapters. So if you skip ahead you may miss important skills taught in the chapters you didn't read. For someone who has no - or minimal - programming experience, I recommend reading each chapter in order. If you have more programming experience and just want to learn how to do a specific thing, feel free to skip directly to that chapter.

Citing this book

If this book was useful in your research, please cite it. To cite this book, please use the below citation:

Kaplan J (2022). *Crime by the Numbers: A Criminologist's Guide to R.* https://crimebythenumbers.com/.

BibTeX format:

```
@Manual{crimebythenumbers,
    title = {Crime by the Numbers: A Criminologist's Guide to R},
    author = {Jacob Kaplan},
    year = {2022},
    url = {https://crimebythenumbers.com/},
}
```

How to contribute to this book

If you have any questions, suggestions (such as a topic to cover), or find any issues, please make a post on the Issues page[6] for this book on GitHub. On this page you can create a new issue (which is basically just a post on this

---

[6]https://github.com/jacobkap/crimebythenumbers/issues

forum) with a title and a longer description of your issue. You'll need a GitHub account to make a post. Posting there lets me track issues and respond to your message or alert you when the issue is closed (i.e. I've finished or denied the request). Issues are also public so you can see if someone has already posted something similar.

For more minor issues like typos or grammar mistakes, you can edit the book directly through its GitHub page. That'll make an update for me to accept, which will change the book to include your edit. To do that, click the edit button at the top of the site - the button is highlighted in the below figure. You will need to make a GitHub account to make edits. When you click on that button you'll be taken to a page that looks like a Word doc where you can make edits. Make any edits you want and then scroll to the bottom of the page. There you can write a short (please, no more than a sentence or two) description of what you've done and then submit the changes for me to review.

Please only use the above two methods to contribute or make suggestions about the book. Don't email me. While it's a bit more work for you to do it this way, since you'll need to make a GitHub account if you don't already have one, it helps me. I wrote this book, in part, to help my career so having evidence that people read it and are contributing to it is important to me. It's a way to publicly measure the book's impact.

Where to find data included in this book

To download the data used in this book please see here.[7] Each of the files that are used in this book are available to download at that link. At the top of every chapter that uses one of these files I'll say exactly which file(s) you need

---

[7] https://github.com/jacobkap/crimebythenumbers/tree/master/data

to download. The best way to use this book is to follow along by downloading the data and running the code that I include in each chapter.

Where to find code included in this book

If you're reading this book through its website,[8] you can easily copy the code by clicking on the "Copy to clipboard" option on the top right of every chunk of code. This button, shown in the image below, will copy all of the code in the chunk and you can then paste (through Control/Command+V) into R.

I've also made each chapter available to download as an R file that has every line of code used in each chapter available to you to run. To download the files, please go to the book's GitHub page here.[9] I've saved each chapter twice - once where it only includes the code used (in the "just_code" folder) and once where it includes the code and all of the text in the chapter (in the "code_and_text" folder). So download whichever one you want to use. The code is identical in each.

---

[8] https://crimebythenumbers.com
[9] https://github.com/jacobkap/crimebythenumbers/tree/master/code_repository

About the author

Jacob Kaplan is a researcher at the Princeton School of Public and International Affairs. He holds a PhD from the University of Pennsylvania.

He is the author of several R packages that make it easier to work with data, including fastDummies[10] and asciiSetupReader.[11] His website[12] allows easy analysis of crime-related data, and he has released over a dozen crime data sets[13] that he has compiled, cleaned, and made available to the public. He is also the author of books on the two primary criminal justice data sets: the FBI's Uniform Crime Reporting (UCR) Program Data[14] and the FBI's National Incident Based Reporting System (NIBRS)[15] data.

---

[10]https://jacobkap.github.io/fastDummies/
[11]https://jacobkap.github.io/asciiSetupReader/
[12]http://jacobdkaplan.com/
[13]http://jacobdkaplan.com/data.html
[14]https://ucrbook.com/
[15]https://nibrsbook.com/

Part I

Introduction

1

A soup to nuts project example

Before we get into exactly how to use R, we'll go over a brief example of a kind of data project that you'd do in the real world. For this chapter we'll look at FBI homicide data that you can download here.[1] The file is called "shr_1976_2020.rds".

1.1   Big picture data example

Below is a large chunk of R code along with some comments about what the code does. The purpose of this example is to show that with relatively little code (excluding blank lines and comments, there are only 35 lines of R code here) you can go from opening a data set to making a graph that answers your research question. I don't expect you to understand any of this code as it is fairly complex and involves many different concepts in programming. So if the code is scary - and for many early programmers seeing a bunch of code that you don't understand is scary and overwhelming - feel free to ignore the code itself.

We'll cover each of these skills in turn throughout the book so that by the end of the book you should be able to come back and understand the code (and modify it to meet your own needs). The important thing is that you can see exactly what R can do (and this is only a tiny example of R's flexibility) and think about the process to get there (which we'll talk about below).

At the time of this writing, the FBI had just released 2020 crime data, which showed about a 30% increase in murders relative to 2019. This had led to an explosion of (in my opinion highly premature) explanations of why exactly murder went up so much in 2020. A common explanation is that it is largely driven by gun violence among gang members who are killing each other in a cyclical pattern of murders followed by retaliatory murders. For our coding example, we'll examine that claim by seeing if gang violence did indeed increase, and whether it increased more than other types of murders.

---

[1] https://github.com/jacobkap/crimebythenumbers/tree/master/data

3

The end result is the graph below. It is, in my opinion, a fairly strong answer to our question. It shows the percent change in murders by the victim-offender relationship from 2019 to 2020. This is using FBI murder data, which technically does have a variable that says if the murder is gang related, but it's a very flawed variable (i.e. vast undercount of gang-related murders) so I prefer to use stranger and acquaintance murders as a rough proxy. And we now have an easy to read graph that shows that while indeed stranger and acquaintance murders did go up a lot, nearly all relationship groups experienced far more murders in 2020 than in 2019. This suggests that there was a broad increase in murder in 2020, and it was not driven merely by an increase in one or a few groups.

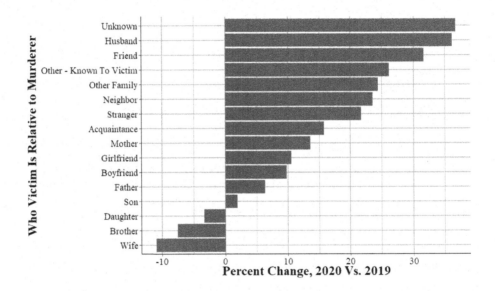

These graphs (though modified to a table instead of a graph) were included in a article I contributed to on the site FiveThirtyEight[2] in discussing the murder increase in 2020. So this is an actual work product that is used in a major media publication - and is something that you'll be able to do by the end of this book. For nearly all research you do you'll follow the same process as in this example: load data into R, clean it somehow, and create a graph or a table or do a regression on it. While this can range from very simple to very complex depending on your exact situation (and how clean the data is that you start with), all research projects are essentially the same.

Please look at the following large chunk of code. We'll next go through each of the different pieces of this code to start understanding how they work.

[2]https://fivethirtyeight.com/features/murders-spiked-in-2020-how-will-that-change-the-politics-of-crime/

Throughout the course of this book we'll cover these steps in more detail - as most research programming work follows the same process - so here we'll talk more abstractly about what each does. The goal is for you to understand the basic steps necessary for using R to do research, and to understand how R can do it - but not having to understand what each line of code does just yet.

```r
library(dplyr) # Used to aggregate data
library(ggplot2) # Used to make the graph
library(crimeutils) # Used to capitalize words in a column
library(tidyr) # Used to reshape the data


# Load in the data
shr <- readRDS("data/shr_1976_2020.rds")


# See which agencies reported in 2019 and 2020
# An "ori" is a unique identifier code for agencies in FBI data
agencies_2019 <- shr$ori[shr$year == 2019]
agencies_2020 <- shr$ori[shr$year == 2020]
# Get which agencies reported in both years so we have an
# apples-to-apples comparison
agencies_in_both <- agencies_2019[agencies_2019 %in% agencies_2020]


# Keep just data from 2019 and 2020 and where the agencies
# is one of the agencies chosen above. Also keep only murder and
# nonnegligent manslaughter (so excluding  negligent manslaughter).
shr_2019_2020 <- shr[shr$year %in% 2019:2020, ]
shr_2019_2020 <- shr_2019_2020[shr_2019_2020$ori %in%
  agencies_in_both, ]
shr_2019_2020 <- shr_2019_2020[shr_2019_2020$homicide_type %in%
  "murder and nonnegligent manslaughter", ]


# Get the number of murders by victim-offender relationship in 2019 and 2020
# Then find the percent change in murders by this group from 2019 to 2020
# Sort data by smallest to largest percent change
shr_difference <-
  shr_2019_2020 %>%
  group_by(year) %>%
  count(victim_1_relation_to_offender_1) %>%
  spread(year, n) %>%
  mutate(
    difference = `2020` - `2019`,
    percent_change = difference / `2019` * 100,
    victim_1_relation_to_offender_1 =
      capitalize_words(victim_1_relation_to_offender_1)
```

```
) %>%
filter(`2019` >= 50) %>%
arrange(percent_change)

# This is only for the graph. By default graphs order alphabetically
# but this makes sure it orders it based on the ordering we made above
# (smallest to largest percent change)
shr_difference$victim_1_relation_to_offender_1 <-
  factor(shr_difference$victim_1_relation_to_offender_1,
    levels = shr_difference$victim_1_relation_to_offender_1
  )

# Makes a barplot showing the percent change from 2019 to 2020 in number
# of murders by victim group. Labels the x-axis and the y-axis, shifts
# the graph so that relationship labels are on the y-axis for easy reading.
# And finally uses the "crim" theme that changes the colors in the graph to
# make it a little easier to see.
ggplot(shr_difference, aes(
  x = victim_1_relation_to_offender_1,
  y = percent_change
)) +
  geom_bar(stat = "identity") +
  ylab("% Change, 2020 Vs. 2019") +
  xlab("Victim Relative to Murderer") +
  coord_flip() +
  theme_crim()
```

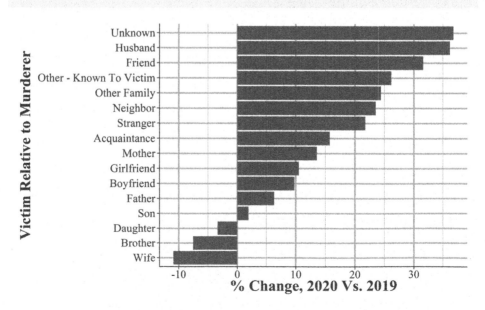

1.2 Little picture data example

We'll now look at each piece of the larger chunk of code above, and I'll explain what it does. There are five different steps that I take to create the graph from the data we use:

1. Load the packages we use
2. Load the data
3. Clean the data
4. Aggregate the data
5. Make the graph

1.2.1 Loading packages

In R we'll often use code written by other people that have tools that we want to use in our code. To use this code we need to tell R that we want to use that particular package - and packages are just a collection of other people's code. A collection of code for a specific purpose (e.g. making a graph, doing a very particular cleaning task) is called a function. Each package is a collection of functions. For this example, we're using packages that help us clean and aggregate data or to graph it, so we load it here. The general convention is to start your R file with each of the packages you want to use at the top of the file.

```
library(dplyr)
library(ggplot2)
library(crimeutils)
library(tidyr)
```

1.2.2 Loading data

Next we need to load in our data. The data we're using is a type of R data file called an .Rds file so we load it using the function readRDS(), which is one of the functions built into R so we don't actually need to use any package for it. For this example, we're using data from the FBI's Supplementary Homicide Report which are an annual data set that has relatively detailed information on most (but not all, as not all agencies report data) murders in the United States. This includes the relationship between the victim and the offender (technically the suspected offender) in the murder, which is what we'll look at. When we read in the data to R we need to give it a name so R knows what it

is called. We'll call this data "shr" since that is the normal abbreviation for
the Supplementary Homicide Report data. Normally in R we use lower cased
letters when naming something, which is why we're calling it "shr" rather
than "SHR."

Each row of data is actually a murder incident, and there can be up to 11
victims per murder incident. So we'll be undercounting murders as in this
example we're only looking at the first victim in an incident. But, as it's an
example, this is fine as I don't want it to be too complicated and including
more than just the first victim would greatly complicate our code.

```
shr <- readRDS("data/shr_1976_2020.rds")
```

1.2.3   Cleaning

One of the annoying quirks of dealing with FBI data is that different agencies
report each year. So comparing different years has an issue because you'll be
doing an apples-to-oranges competition as an agency may report one year but
not another. So for this data the first thing we need to do is to make sure
we're only looking at agencies that reported data in both years. The first few
lines check which agencies reported in 2019 and which agencies reported in
2020. We do this by looking at which ORIs (in the "ori" column) are present
in each year (as agencies that did not report won't be in the data). An ORI is
the FBI term for a unique ID for that agency. Then we make a vector, which
has only the ORIs that are present in both years.

We then subset the data to only data from 2019 and 2020 and where the
agency reported in both years. Subsetting essentially means that we only keep
the rows of data that meet those conditions. Another quirk of this data is that
it includes homicides that are not murder - namely, negligent manslaughter.
So the final subsetting condition we use is that it only includes murder and
nonnegligent manslaughter.

```
agencies_2019 <- shr$ori[shr$year == 2019]
agencies_2020 <- shr$ori[shr$year == 2020]
agencies_in_both <- agencies_2019[agencies_2019 %in% agencies_2020]


shr_2019_2020 <- shr[shr$year %in% 2019:2020, ]
shr_2019_2020 <- shr_2019_2020[shr_2019_2020$ori %in% agencies_in_both, ]
shr_2019_2020 <- shr_2019_2020[shr_2019_2020$homicide_type %in%
  "murder and nonnegligent manslaughter", ]
```

1.2.4 Aggregating

Now we have only the rows of data that we want. Each row of data is a single murder incident, so we want to aggregate that data to the year-level and see how many murders there were for each victim-offender relationship group. The following chunk of code does that and then finds the percent difference. Since we can have large percent changes due to low base rates, we then remove any rows where there were fewer than 50 murders of that victim-offender relationship type in 2019. Finally, we arrange the data from smallest to largest difference. We'll print out the data just to show you what it looks like.

```
shr_difference <-
  shr_2019_2020 %>%
  group_by(year) %>%
  count(victim_1_relation_to_offender_1) %>%
  spread(year, n) %>%
  mutate(
    difference = `2020` - `2019`,
    percent_change = difference / `2019` * 100,
    victim_1_relation_to_offender_1 =
      capitalize_words(victim_1_relation_to_offender_1)
  ) %>%
  filter(`2019` >= 50) %>%
  arrange(percent_change)
shr_difference
# # A tibble: 16 x 5
#    victim_1_relation_to_offe~1 `2019` `2020` diffe~2 perce~3
#    <chr>                        <int>  <int>   <int>   <dbl>
#  1 Wife                           330    294     -36   -10.9
#  2 Brother                         93     86      -7   -7.53
#  3 Daughter                        89     86      -3   -3.37
#  4 Son                            157    160       3    1.91
#  5 Father                          95    101       6    6.32
#  6 Boyfriend                      164    180      16    9.76
#  7 Girlfriend                     390    431      41    10.5
#  8 Mother                         118    134      16    13.6
#  9 Acquaintance                  1494   1729     235    15.7
# 10 Stranger                      1549   1886     337    21.8
# 11 Neighbor                        85    105      20    23.5
# 12 Other Family                   209    260      51    24.4
# 13 Other - Known To Victim        757    955     198    26.2
# 14 Friend                         272    358      86    31.6
# 15 Husband                         58     79      21    36.2
```

```
# 16 Unknown                      6216    8504    2288    36.8
# # ... with abbreviated variable names
# #   1: victim_1_relation_to_offender_1, 2: difference,
# #   3: percent_change
```

1.2.5   Graphing

Once we have our data cleaned and organized in the way we want, we are ready to graph it. By default when R graphs data it will organize it alphabetically. In our case we want it ordered by smallest to largest change in the number of murders between 2019 and 2020 by relationship type. So we first tell R to order it by the relationship type variable, which we've already sorted in the last section of code. Then we use the ggplot() function (which is covered extensively in Chapters 14 and 15) to make our graph. In our code we include the data set we're using, which is the shr_difference data and the columns we want to graph. Then we tell it we want to create a bar chart and what we want the x-axis and y-axis labels to be. Finally, we have two lines that just affect how the graph looks. All of this is covered in the two graphing chapters, but is only several lines of code to go from cleaned data to a beautiful - and informative - graphic.

```
shr_difference$victim_1_relation_to_offender_1 <-
  factor(shr_difference$victim_1_relation_to_offender_1,
    levels = shr_difference$victim_1_relation_to_offender_1
  )

ggplot(shr_difference, aes(
  x = victim_1_relation_to_offender_1,
  y = percent_change
)) +
  geom_bar(stat = "identity") +
  ylab("% Change, 2020 Vs. 2019") +
  xlab("Victim Relative to Murderer") +
  coord_flip() +
  theme_crim()
```

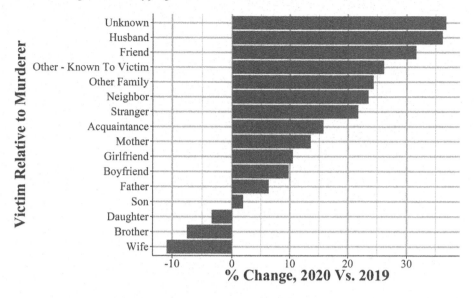

1.3 Reusing and modifying code

One of the main benefits of programming is that once you write code to do one thing, it's usually very easy to adapt it to do a similar thing. Below I've copied some of the code we used above and changed only one thing: instead of looking at the column "victim_1_relation_to_offender_1" we're now looking at the column "offender_1_weapon". That's all I did, everything else is identical. Now after about 30 seconds of copying and changing the column name, we have a graph that shows weapon usage changes from 2019 to 2020 instead of victim-offender relationship.

This is one of the key benefits of programming over something more click intensive like using Excel or SPSS.[3] There's certainly more upfront work than just clicking buttons, but once we have working code we can very quickly reuse it or modify it slightly.

```
shr_difference <-
  shr_2019_2020 %>%
  group_by(year) %>%
  count(offender_1_weapon) %>%
```

---

[3]I'm aware that technically you can write SPSS code. However, every single person I know who has ever used SPSS does so by clicking buttons and is afraid of writing code.

```
  spread(year, n) %>%
  mutate(
    difference = `2020` - `2019`,
    percent_change = difference / `2019` * 100,
    offender_1_weapon = capitalize_words(offender_1_weapon)
  ) %>%
  filter(`2019` >= 50) %>%
  arrange(percent_change)

shr_difference$offender_1_weapon <-
  factor(shr_difference$offender_1_weapon,
    levels = shr_difference$offender_1_weapon
  )
ggplot(shr_difference, aes(
  x = offender_1_weapon,
  y = percent_change
)) +
  geom_bar(stat = "identity") +
  ylab("% Change, 2020 Vs. 2019") +
  xlab("Offender Weapon") +
  coord_flip() +
  theme_crim()
```

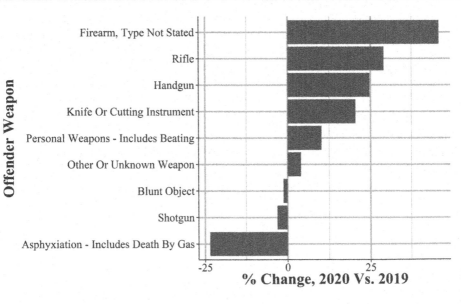

2

Introduction to R and RStudio

In this chapter you'll learn to open a data file in R. That file is "ucr2017.rda," which you'll need to download from the data repository available here.[1]

2.1 Using RStudio

In this lesson we'll start by looking at RStudio then write some brief code to load in some crime data and start exploring it. This lesson will cover code that you won't understand completely yet. That is fine, we'll cover everything in more detail as the lessons progress.

RStudio is the interface we use to work with R. It has a number of features to make it easier for us to work with R. While not strictly necessary to use, most people who use R do so through RStudio. When using R you don't need to open up both R and RStudio on your computer. Just open RStudio and it'll internally use R. We'll spend some time right now looking at RStudio and the options you can change to make it easier to use (and to suit your personal preferences with appearance) as this will make all of the work that we do in this book easier.

When you open up RStudio you'll see four panels, each of which plays an important role in RStudio. Your RStudio may not look like the setup I have in the following image - that is fine, we'll learn how to change the appearance of RStudio soon.

---

[1]https://github.com/jacobkap/crimebythenumbers/tree/master/data

At the top right of the image (and this may be in a different location on your RStudio) is the Console panel. Here you can write code, hit enter/return, and R will run that code. If you write 2+2 it will return (in this case that just mean it will print an answer) 4. This is useful for doing something simple like using R as a calculator or quickly looking at data. In most cases during research this is where you'd do something that you don't care to keep. This is because when you restart R it won't save anything written in the Console. To do reproducible research or to be able to collaborate with others you need a way to keep the code you've written.

The way to keep the code you've written in a file that you can open later or share with someone else is by writing code in an R Script (if you're familiar with Stata, an R Script is just like a .do file). An R Script is essentially a text file (similar to a Word document) where you write code. To run code in an R Script just click on a line of code or highlight several lines and hit enter/return or click the "Run" button on the top right of the Source panel shown in the top left of the above image. You'll see the lines of code run in the Console and any output (if your code has an output) will be shown there too (making a plot will be shown in a different panel as we'll see soon).

For code that you don't want to run, called comments, start the line with a pound sign # and that line will not be run (it will still print in the console if you run it but it won't do anything). These comments should explain the code you wrote (if it's not otherwise obvious what the code does).

It is good practice to do all of your code writing in an R Script - even if you delete some lines of code later - as it eliminates the possibility of losing code or forgetting what you wrote. Having all the code in front of you in a text file also makes it easier to understand the flow of code from start to finish for a task - an issue we'll discuss more in later lessons.

While the Source and Console panels are the ones that are of most use, there are two other panels worth discussing. As these two panels let you interchange which tabs are available in them, we'll return to them shortly in the discussion of the options RStudio has to customize it.

2.1.1 Opening an R Script

When you want to open up a new R Script you can click File on the very top left, then R Script. It will open up the script in a new tab inside of the Source panel. There are also a number of other file options available: R Presentation which can make PowerPoints; R Markdown, which can make Word Documents or PDFs that incorporate R code used to make tables or graphs (and which we'll cover in Chapter 7); and Shiny Web App to make websites using R. There is too much to cover for an introductory book such as this, but keep in mind the wide capabilities of R if you have another task to do. To open an R Script that is already saved to your computer, click "Open File..." and navigate to the file that you want to open.

2.1.2   Setting the working directory

Many research projects incorporate data that someone else (such as the FBI or a local police agency) has put together. In these cases, we need to load the data into R to be able to use it. In a little bit we'll load a data set into R and start working on it, but let's take a step back now and think about how to even load data. First, we'll need to get the data onto our computer somehow, probably by downloading it from an agency's website. Let's be specific - we don't download it to our computer, we download it to a specific folder on our computer (usually defaulted to the Downloads folder on a Windows machine). So let's say you wanted to load a file called "data" into R. If you have a file called "data" in both your Desktop and your Downloads folder, R wouldn't know which one you wanted. And unless your data was in the folder R searches by default (which may not be where the file is downloaded by default), R won't know which file to load.

We need to tell R explicitly which folder has the data to load. We do this by setting the "Working Directory" (or the "Folders where I want you, R, to look for my data" in more simple terms). To set a working directory in R click the Session tab on the top menu, scroll to Set Working Directory, then click Choose Directory. This will open a window where you can navigate to the folder you want.

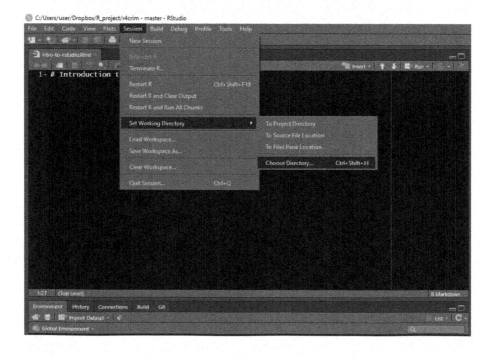

After clicking Open in that window you'll see a new line of code in the Console starting with `setwd()` and inside of the parentheses is the route your computer takes to get to the folder you selected. And now R knows which folder to look in for the data you want. It is good form to start your R Script with `setwd()` to make sure you can load the data. Copy the line of code that says `setwd()` (which stands for "set working directory"), including everything in the parentheses, to your R Script when you start working.

2.1.3 Changing RStudio

Your RStudio looks different than my RStudio because I changed a number of settings to suit my preferences. To do so yourself click the Tools tab on the top menu and then click Global Options.

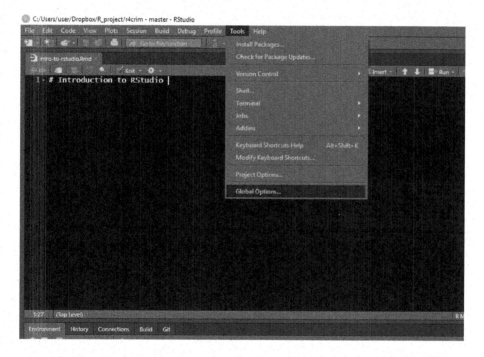

This opens up a window with a number of different tabs to change how R behaves and how it looks.

2.1.3.1 General

Under Workspace in the General tab make sure to **uncheck** the "Restore .RData into workspace at startup" and to set "Save workspace to .RData on exit:" to **Never**. What this does is make sure that every time you open

RStudio it starts fresh with no objects (essentially data loaded into R or made in R) from previous sessions. This may be annoying at times, especially when it comes to loading large files, but the benefits far outweigh the costs.

You want your code to run from start to finish without any errors. Something I've seen many students do is write some code in the Console (or in their R Script but out of order of how it should be run) to fix an issue with the data. This means their data is how it should be, but when the R session restarts (such as if the computer restarts) they won't be able to get back to that point. Making sure your code handles everything from start to finish is well-worth the avoided headache of trying to remember what code you did to fix the issue previously.

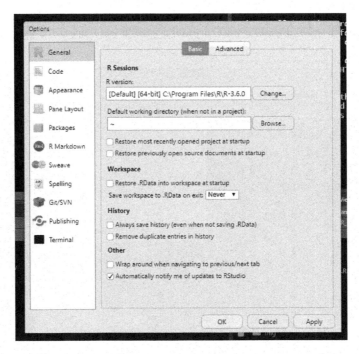

2.1.3.2 Code

The Code tab lets you specify how you want the code to be displayed. The important section for us is to make sure to check the "Soft-wrap R source files" check-box. If you write a very long line of code it gets too big to view all at once and you must scroll to the right to read it all. That can be annoying as you won't be able to see all the code at once. Setting "Soft-wrap" makes it so if a line is too long it will just be shown on multiple lines, which solves that issue. In practice it is best to avoid long lines of codes as it makes it hard to read, but that isn't always possible.

2.1.3.2.1 Saving

Inside of the Code tab we also want to turn on an option to have RStudio automatically save the R script when we aren't using it. This is like how Google Docs automatically saves your document every second or so. While we should be saving our file often (using the little floppy disk icon near the top of RStudio), having RStudio automatically save adds a level of security as it prevents losing a lot of progress if we forget to save and RStudio crashes or we close it.

To set it to autosave, move to the Saving tab, and check the "Automatically save when editor loses focus" box. So if you click out of RStudio or stop typing, it will automatically save. You can also say how long to wait before saving with options ranging from 500 milliseconds to 10,000 milliseconds, which is the same as 0.5 seconds to 10 seconds.

2.1.3.3   Appearance

The Appearance tab lets you change the background, color, and size of text.
Change it to your preferences.

2.1.3.4 Pane Layout

The final tab we'll look at is Pane Layout. This lets you move around the Source, Console, and the other two panels. There are a number of different tabs to select for the panels (unchecking one just moves it to the other panel, it doesn't remove it from RStudio), and we'll talk about three of them. The Environment tab shows every object you load into R or make in R. So if you load a file called "data" you can check the Environment tab. If it is there, you have loaded the file correctly.

As we'll discuss more in Section 2.3, the Help tab will open up to show you a help page for a function you want more information on (we'll also discuss exactly what a function is below. But for now just think of a function as a shortcut to using code that someone else wrote). The Plots tab will display any plot you make. It also keeps all plots you've made (until restarting RStudio) so you can scroll through the plots.

2.1.4   Helpful cheat sheets

RStudio also includes a number of links to helpful cheat sheets for a few important topics. To get to it click Help, then Cheatsheets, and click on whichever one you need.

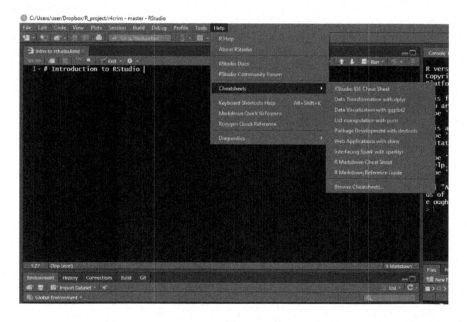

2.2 Assigning variables

When we're using R for research the general process is to load data, change it somehow (such as deleting rows we don't want, aggregating from some small unit such as monthly crime to a higher unit such as yearly crime), and then analyze it. To do all this we need to be able to make sure each step we do actually changes the data. This seems simple but is actually a very common issue I've noticed when working with new R programmers - they run code on the data (e.g. deleting certain rows) but forget to save the change to that data.

Let's look at an example of this. First, we need to know how to create objects in R. I use "object" in a very vague sense to mean anything that is loaded into R and can be manipulated. To create something in R we assign "something" to an object name. This is a very technical sentence so let's look at an example and then step back and try to understand that sentence.

```
a <- 1
```

Above I am creating the object "a" by assigning it the value of 1. In R terms, "a is assigned 1" or "a gets 1". In non-technical terms: a equals 1.

We can print out a to see if this is true.

```
a
# [1] 1
```

When we print out a, it returns 1 since that was what a was assigned to. We can assign a another value, and it will overwrite 1 with whatever value we choose.

```
a <- 33
a
# [1] 33
```

Now a is 33. Or a equals 33. Or a was assigned 33. Or a gets 33. Or we assigned 33 to a. There are a lot of ways to explain what we did here, which is quite frustrating and confusing to new R programmers. I use the terms "assignment" and "gets" only because that is the convention in R, but if it's easier for you to talk about something equaling something else (instead of being assigned to that value), please do so!

The <- is what does the assignment, or what makes the thing on the left equal to the thing on the right. You might be thinking that it'd be easier to simply use the equal sign instead of the <- - we are making things equal after all. And you'd be right. Using = does the exact same thing as <-.

```
a <- 13
a
# [1] 13
```

We can use = instead of <- and get the same results (with very few exceptions and none that are relevant in this book). The reason that people use <- instead of = is largely a matter of convention. It's just the thing that R programmers do so new programmers tend to adopt it. If it's easier for you to use = instead of <-, feel free to do that.

In this book I'll use <- and talk about "assigning" values because that is the convention in R. And while that's not really a good reason to do anything, I think that it's important that new R programmers at least know what the proper conventions are and be able to speak the language (so to speak) of R programmers. This is also important when searching for more help on a topic as you need to know the right term to be able to ask for help (from other R programmers and from Google) easily.

So far we've just been assigning "a" a value, or overwriting that value with a new value. We can also assign something new to have the same value as a. Let's make the object "example_123_value.demonstration" get the value that a has - or in other words make "example_123_value.demonstration" be equal to a.

```
example_123_value.demonstration <- a
example_123_value.demonstration
# [1] 13
```

I use name "example_123_value.demonstration" just an example of what you can include in an object name - any character (lower or uppercase), any number (just can't start with a number), and some punctuation (e.g. underscores and periods). Spaces are not allowed. In practice you'll want to call each object something specific so you know what it is, and ideally make the name as short as possible. For example, if you are using crime data from Houston you'll want to call it something like "houston_crime". The R convention is to only use lowercase characters and include only underscores as the punctuation, but you can name it whatever is most useful to you.

As noted at the start of the section, a lot of new programmers will make a change to an object but forget to assign the result back into the object (or into a new object). This means that that object won't actually change. For example, let's say we want to multiply example_123_value.demonstration by 10.

If we do `example_123_value.demonstration * 10` then it'll print out the result in the console, but not actually change example_123_value.demonstration. What we need to do is assign that result of the multiplication back into example_123_value.demonstration. Lots of new programmers forget to assign the results back into the object, which understandably leads to lots of confusion since the object is now not what they expect it to be.

```
example_123_value.demonstration <- example_123_value.demonstration * 10
example_123_value.demonstration
# [1] 130
```

I've been saying "object" a lot, without defining it. An object is a bit tricky to define, especially at this stage in the book. Throughout this book I'll be using object to describe something that has been assigned value, such as "a" and "example_123_value.demonstration". This also includes outside data sets read into R, such as an Excel file loaded into R and even a set of R code that has been assigned to an object (which is called a function). Each object that you have created or loaded yourself can be found in the Environment tab.

2.3 What are functions (and packages)?

When programming to do research you'll often have to do the same thing multiple times. For example, many crime data sets are available as one file for each year of data. So if you are analyzing multiple years of data you'll need to clean each file separately - and in most cases that involves using the exact same code for every file. This also includes doing things that other people have done. For example, most research leads to at least one graph being made. Since making graphs is so common, many people have spent a long time writing code to make it easy to make publication-ready graphs. Instead of doing all that work ourselves we can just use code that other people have written and made available to us. While we could do this by copying code, the easiest way to reuse code is to use functions.

As noted in the previous section, a function is a bunch of code (it could range from a single line of code to hundreds of lines) that has been assigned to an object. We'll dive into this topic in detail in Chapter 20 - including how to make your own functions - but using functions is such an important concept that we'll briefly introduce them here. Almost everything that you will do in R is through functions. For the most part that'll be using functions that other people have written that are available to use - and this includes functions that are built into R already and ones we have to download from other R programmers.

Let's look at the function `head()` as an example. This is a function that is already built into R which means we don't need to do anything to use it. For functions that are written by other R programmers we'll need to download those functions and tell R we want to use it - and we'll show how in a bit. The way to identify a function is through the parentheses after the function name (the naming convention is the same as for objects as discussed in the previous section. We want a short, descriptive name that explains what the function does). If we see a word followed by parentheses, we can be confident that we're looking at a function.

The `head()` function prints out the first 6 rows of every column of a data.frame (which is essentially an Excel sheet, and something we'll cover in more detail in Chapter 3). `head()` is an extremely useful and common function in R, but just the name alone doesn't make it clear what it does or that we need to put a data object inside the parentheses.

If you are having trouble understanding what a function does or how to use it, you can ask R for help and it will open up a page explaining what the function does, what options it has, and examples of how to use it. To do so we write `help(function)` or `?function` in the console and it will open up that function's help page. For finding the help page of a function we do not include the parentheses part of the function: `help(head)` works while `help(head())` does not.

If we wrote `help(head)` to figure out what the `head()` function does, it will open up this page. Unfortunately, many help pages are not that useful. The following image shows the help page for `head()`, and it is not very friendly to a new R programmer. In cases where the help page is not useful, and you're looking at functions not covered in this book, I recommend looking online for help pages dedicated to that function or broader programming sites such as Stack Overflow,[2] where people can ask questions about programming.

---

[2]https://stackoverflow.com/

For `head()`, all we need to do is tell the function what data we're looking at. In programming terms, the input to the function (what we have to include in the parentheses) is the name of our data object. We'll look at the very commonly used data called `mtcars`. `mtcars` is one of a small number of data files that are already in R when you open it. These are included in R just as examples of data to use when testing our code or teaching people to use R. Just type `mtcars` into the console and it will print out data to the console; there's nothing you need to do to load the data into R. `mtcars` has info about a number of cars with each row being a type of car and each column being information about the car such as the miles per gallon it gets and how many gears it has.

We'll use the `head()` function to print out just the first 6 rows of the `mtcars` data.

```
head(mtcars)
#                     mpg cyl disp  hp drat    wt  qsec vs am
# Mazda RX4          21.0   6  160 110 3.90 2.620 16.46  0  1
# Mazda RX4 Wag      21.0   6  160 110 3.90 2.875 17.02  0  1
# Datsun 710         22.8   4  108  93 3.85 2.320 18.61  1  1
# Hornet 4 Drive     21.4   6  258 110 3.08 3.215 19.44  1  0
# Hornet Sportabout  18.7   8  360 175 3.15 3.440 17.02  0  0
# Valiant            18.1   6  225 105 2.76 3.460 20.22  1  0
#                   gear carb
# Mazda RX4            4    4
```

```
# Mazda RX4 Wag        4    4
# Datsun 710           4    1
# Hornet 4 Drive       3    1
# Hornet Sportabout    3    2
# Valiant              3    1
```

Now we have the first 6 rows of every column from the mtcars data. This is a fairly simple function and is useful for quickly looking at our data. Many functions are more complicated than head() and involve multiple inputs rather than just the single input we had here. Some functions, for example, let you choose how you want the function to operate, as it can do so in multiple ways. Even in head() there's an optional input to choose how many rows you want it to return, with the default being 6. Since we didn't choose anything, the function stuck to the default and returned only 6 rows.

Throughout this book we'll spend a lot of time introducing functions that other people have made and learning how to combine the functions together to be able to get our raw data (e.g. a CSV file downloaded from a police site) into a usable format for research (e.g. cleaned to include only the rows and columns we need to analyze and in the units we want). For functions that other people wrote, we need to tell R that we want to use these functions. We do so by having R download that person's package. A package is just the name for a collection of functions in an easily downloadable format. We can do all of the downloading through R, so we don't have to go searching for them. There are two ways to download a package in R: through writing R code or through a shortcut in RStudio.

Downloading a package through R code uses - like pretty much everything else in R - a function. This function is install.packages(), where we put the name of the package we want in the (). This name also has to be in quotes since it is an object that is not currently in R. Let's install the package "caesar", which is a simple package I made that creates a Caesar cipher from some text. We need to run the code install.packages("caesar") and be sure to spell "caesar" right and put it in quotes.

```
install.packages("caesar")
```

The RStudio shortcut way is to go to the Packages tab and then click Install on the top left of this tab. This will open up a window as shown in the following image where you can enter the name of the package you want. Then click Install and RStudio will install it for you. Also in this tab is the Update button, which allows you to update packages that you have already installed. Since R programmers generally provide updates to their packages (usually bug fixes but occasionally new features and new functions), it's important to update your packages every several months or so.

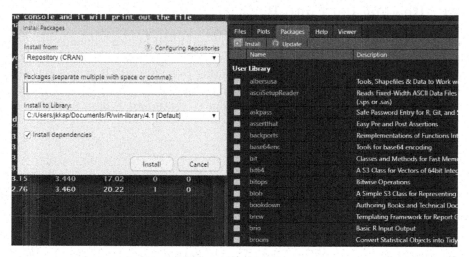

Once we have downloaded the package, we need to tell R that we want to use that package. There are thousands of R packages and you'll likely have hundreds downloaded before long (if a package relies on other packages to work it'll download those too. So even if you install a single package it may also install other packages necessary for the package you want). Some packages have functions with the same name (but they do different things) so using all packages at once will cause issues since we won't know which functions we're actually using. So we only want to use the packages we need for that task. We need a way to tell R that we want to use a package. We only need to do this once per session - that is, once before restarting RStudio. The way to do this is to use the function `library()`, where we put the package name in the parentheses. Since the package is something that has been installed to R, we don't need to have quotes around the name.

```
library(caesar)
```

Now we can run the `caesar()` function and make a Caesar cipher for that text (it's just a coincidence that the function name is the same as the package name).

```
caesar("example text")
# [1] "hAdpsohcwhAw"
```

2.4   Reading data into R

For many research projects you'll have data produced by some outside group (e.g. FBI, local police agencies) and you want to take that data and put it inside R to work on it. We call that reading data into R. R is capable of reading a number of different formats of data, which we will discuss in more detail in Chapter 4. Here, we will talk about the standard R data file only.

2.4.1   Loading data

As we learned in Section 2.1.2, we need to set our working directory to the folder where the data is. For my own setup, R is already defaulted to the folder with this data so I do not need to set a working directory. For those following along on your own computer, make sure to set your working directory now.

The `load()` function lets us load data already in the R format. These files will end in the extension ".rda" or sometimes ".Rda" or ".RData". Since we are telling R to load a specific file, we need to have that file name in quotes and include the file extension ".rda". With .rda data, the object inside the .rda file already has a name so we don't need to assign a name to the data. With other forms of data such as .csv files, we will need to do that as we'll see in Chapter 4.

In this example (and elsewhere in this book when I load in data), I have all of the data in a folder called "data" in my working directory, which is why I have "data/" before the data name. You do not need this as you should have all of your data directly in your working directory.

```
load("data/ucr2017.rda")
```

2.5   First steps to exploring data

The object we loaded is called `ucr2017`. We'll explore this data more thoroughly in Chapter 11, but for now let's use four simple (and important) functions to get a sense of what the data holds. To use each of these functions, we need to write the name of the data set (without quotes since we don't need quotes for an object already made in R) inside the ().

- `head()`
- `summary()`
- `plot()`
- `View()`

Note that the first three functions are lowercase while `View()` is capitalized. That is simply because older functions in R were often capitalized while newer ones use all lowercase letters. R is case sensitive so using `view()` will not work.

The `head()` function prints the first 6 rows of each column of the data to the console. This is useful to get a quick glance at the data but has some important drawbacks. When using data with a large number of columns it can be quickly overwhelming by printing too much. There may also be differences in the first 6 rows with other rows. For example, if the rows are ordered chronologically (as is the case with most crime data) the first 6 rows will be the most recent. If data collection methods or the quality of collection changed over time, these 6 rows won't be representative of the data.

```
head(ucr2017)
#         ori year agency_name   state population actual_murder
# 1 AK00101 2017   anchorage  alaska     296188            27
# 2 AK00102 2017   fairbanks  alaska      32937            10
# 3 AK00103 2017      juneau  alaska      32344             1
# 4 AK00104 2017   ketchikan  alaska       8230             1
# 5 AK00105 2017      kodiak  alaska       6198             0
# 6 AK00106 2017        nome  alaska       3829             0
#   actual_rape_total actual_robbery_total
# 1               391                  778
# 2                24                   40
# 3                50                   46
# 4                19                    0
# 5                15                    4
# 6                 7                    0
#   actual_assault_aggravated
# 1                      2368
# 2                       131
# 3                       206
# 4                        14
# 5                        41
# 6                        52
```

The `summary()` function gives a six-number summary of each numeric or Date column in the data. For other types of data, such as "character" types (which are just columns with words rather than numbers or dates), it'll say what type of data it is. We'll cover different types of data in Chapter 3.

The six values it returns for numeric and Date columns are

- The minimum value
- The value at the 1st quartile
- The median value
- The mean value
- The value at the 3rd quartile
- The max value

In cases where there are NAs, it will say how many NAs there are. An NA value is a missing value. Think of it like an empty cell in an Excel file. NA values will cause issues when doing math, such as finding the mean of a column, as R doesn't know how to handle a NA value in these situations, though summary() automatically excludes NAs when doing the math operations.

```
summary(ucr2017)
#       ori                   year          agency_name
#   Length:15764       Min.    :2017     Length:15764
#   Class :character   1st Qu.:2017     Class :character
#   Mode  :character   Median :2017     Mode  :character
#                      Mean    :2017
#                      3rd Qu.:2017
#                      Max.    :2017
#       state              population      actual_murder
#   Length:15764       Min.    :      0   Min.    :  0.000
#   Class :character   1st Qu.:    914   1st Qu.:  0.000
#   Mode  :character   Median :   4460   Median :  0.000
#                      Mean    :  19872   Mean    :  1.069
#                      3rd Qu.:  15390   3rd Qu.:  0.000
#                      Max.    :8616333   Max.    :653.000
#   actual_rape_total  actual_robbery_total
#   Min.    :  -2.000  Min.    :   -1.00
#   1st Qu.:   0.000  1st Qu.:    0.00
#   Median :   1.000  Median :    0.00
#   Mean    :   8.262  Mean    :   19.85
#   3rd Qu.:   5.000  3rd Qu.:    4.00
#   Max.    :2455.000  Max.    :13995.00
#   actual_assault_aggravated
#   Min.    :   -1.00
#   1st Qu.:    1.00
#   Median :    5.00
#   Mean    :   49.98
#   3rd Qu.:   21.00
#   Max.    :29771.00
```

The `plot()` function allows us to graph our data. For criminology research we generally want to make scatterplots to show the relationship between two numeric variables, time-series graphs to see how a variable (or variables) change over time, or barplots comparing categorical variables. Here, we'll make a scatterplot seeing the relationship between a city's number of murders and their number of aggravated assaults (assault with a weapon or that causes serious bodily injury).

To do so we must specify which column is displayed on the x-axis and which one is displayed on the y-axis. In Section 10.3.1 we'll talk explicitly about how to select specific columns from our data. For now, all you need to know is to select a column in which you write the data set name followed by a dollar sign $, followed by the column name. Do not include any quotations or spaces (technically spaces can be included but make it a bit harder to read and are against conventional style when writing R code so we'll exclude them). Inside of `plot()` we say that "x = ucr2017$actual_murder" so that column goes on the x-axis and "y = ucr2017$actual_assault_aggravated" so aggravated assault goes on the y-axis. And that's all it takes to make a simple graph.

```
plot(x = ucr2017$actual_murder, y = ucr2017$actual_assault_aggravated)
```

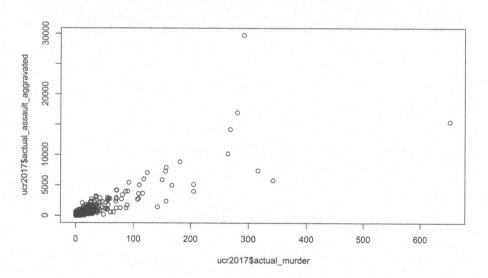

Finally, `View()` opens essentially an Excel file of the data set you put inside the (). This allows you to look at the data as if it were in Excel (though you can't edit the data at all here) and is a good way to start to understand the data.

```
View(ucr2017)
```

	ori	year	agency_name	state	population	actual_murder	actual_rape_total	actual_robbery_total	actual_assault_aggravat
1	AK00101	2017	anchorage	alaska	296188	27	391	778	
2	AK00102	2017	fairbanks	alaska	32937	10	24	40	
3	AK00103	2017	juneau	alaska	32344	1	50	46	
4	AK00104	2017	ketchikan	alaska	8230	1	19	0	
5	AK00105	2017	kodiak	alaska	6198	0	15	4	
6	AK00106	2017	nome	alaska	3829	0	7	0	
7	AK00107	2017	petersburg	alaska	3108	0	2	1	
8	AK00108	2017	seward	alaska	2806	0	0	0	
9	AK00109	2017	sitka	alaska	8736	1	0	1	
10	AK00110	2017	skagway	alaska	1100	0	0	0	
11	AK00111	2017	wrangell	alaska	2395	0	2	0	
12	AK00112	2017	valdez	alaska	3842	0	5	0	
13	AK00113	2017	bethel	alaska	6426	1	31	14	
14	AK00114	2017	cordova	alaska	2199	0	0	0	
15	AK00116	2017	kotzebue	alaska	3251	0	16	3	
16	AK00117	2017	palmer	alaska	7184	1	1	4	
17	AK00118	2017	north slope borough	alaska	9539	2	15	1	
18	AK00120	2017	soldotna	alaska	4693	0	16	6	
19	AK00121	2017	haines	alaska	2472	0	3	0	
20	AK00122	2017	homer	alaska	5735	0	5	1	
21	AK00123	2017	kenai	alaska	7852	0	16	1	
22	AK00126	2017	fairbanks intrnl airport	alaska	0	0	0	0	
23	AK00127	2017	t stevens anchrge int ap	alaska	0	0	0	1	
24	AK00130	2017	dillingham	alaska	2369	0	10	3	
25	AK00132	2017	north pole	alaska	2248	0	1	1	
26	AK00133	2017	unalaska	alaska	4449	0	0	0	

3

Data types and structures

3.1 Data types

When you read a sentence like "two plus two" you know the answer is four. R doesn't know that. This is because R takes things very literally. It will read "two" as a word, not as a number. For R to understand numbers you need to specify that you're talking about numbers, and not just words. Let's look at an example, making two variables which each have the value of "2."

```
a <- "2"
b <- "2"
```

We now have a and b that are equal to "2" (in quotes!). Let's try to add them.

```
a + b
# Error in a + b: non-numeric argument to binary operator
```

We get an error that is a technical way of saying that we did math on something that isn't a number. That's because we made a and b get "2" with quotes around it, which R interpreted as a word, not as a number. If we change a and b to 2 (without quotes), then R will know that the 2 is a number, and will do math on it.

```
a <- 2
b <- 2
a + b
# [1] 4
```

This may seem like a pretty simple concept but is fundamental to how R works, and can trip up new and experienced programmers alike. R trusts you. It only knows what you tell it. If you tell it that something is a word (by including quotes), it will treat it as a word, even if it looks to you like a number. So we must be very precise about what code we write, as R won't (for the most part) fix our mistakes - though it will give us an error if we try to do something it doesn't like, like add two words.

3.2   Numeric, character, and logical (boolean)

There are three main data types that are important to know for using R to do research: numeric, character, and logical.

A numeric type is a number, and this includes both integers like 2 and decimals like 2.5. You can tell something is numeric if it is a number and there are no quotes around it. 2 is a number, "2" is not. For real data this will likely be something like the age of an individual or the number of crimes in a city. We want it as numeric type because we can do math on numbers. For example, we can find the average age of victims of crimes, or the median number of crimes in a city each week. This won't work unless R knows that these values are numbers.

A character is just a word or a set of words. If it is in quotes it's a character. Other programming languages generally call this a string instead of a character, but they mean the same thing. Pretty much anything that you'd write in English class fits in here.

Finally, a logical data type is just a true or false value, though in R it must be written all in capital letters: TRUE or FALSE. This is also referred to as a Boolean value. Booleans or logical data are useful when comparing two things. For example, we can see if 2 is equal to 3.

```
2 == 3
# [1] FALSE
```

It's not, so R returned FALSE (the == just compares the thing on the left to the thing on the right). This is very useful when we want to keep only certain rows in our data. For example, if we had data on multiple years of crime and we only wanted to keep a single year (let's say 2020), we could tell R to keep only rows where the year equals 2020 - where it is TRUE that that row's year column is equal to what year we want. We'll cover this in great detail in Chapter 10.

While you could try to figure out what type of data something is just by looking at it, R has a number of functions to check for you. We'll look at a few general functions that tell you the type of data something is, and then ones that check if the data is a specific type.

First, the is() function tells you all of the types of data something is - and a value can actually have multiple types. While it can't be both, for example, numeric and character, it can have other data types that we'll look at in the next section. First, let's look at what is() returns (prints out to the console) for a few simple examples.

```
is(2)
# [1] "numeric" "vector"
```

Checking what 2 is tells us that it is both a "numeric" type and a "vector" type.

```
is("2")
# [1] "character"              "vector"
# [3] "data.frameRowLabels" "SuperClassMethod"
```

Checking "2" (in quotes), gives us four different types of data for this value: "character", "vector", "data.frameRowLabels", and "SuperClassMethod". You can ignore the last two types, we just are interested in that it is a "character" type and, like the type of 2, is a "vector".

```
is(TRUE)
# [1] "logical" "vector"
```

Finally, checking what TRUE is returns both "logical" and "vector". We expected logical since TRUE is a logical type. Again, we see that it is also a vector type. TRUE has to be both in capital letters and not be in quotes. If we write it in quotes then R will think it is a character, and if we have it lowercase and without quotes R will think that it is an object (such as something we make using <- and not a Boolean).

```
is("TRUE")
# [1] "character"              "vector"
# [3] "data.frameRowLabels" "SuperClassMethod"
```

```
is(true)
# Error in is(true): object 'true' not found
```

All three of the values we checked say that they are a "vector" type. We'll cover vectors in the next section, but for now let's see one other function that tells us the type of data something is. If we use `class()` instead of `is()` we'll get just the first value returned in the types of data that we input.

```
class(2)
# [1] "numeric"
class("2")
# [1] "character"
class(TRUE)
# [1] "logical"
```

In a lot of cases we'll want to check if some data is a specific type. For example, we might want to check that the year column of a data set is numeric, rather than say character. We do this with three functions, each of which checks that the data input (the data put in the parentheses of the function) is that type of data or not. These functions are: `is.numeric()`, `is.character()`, and `is.logical()`.

Running any of these functions will actually return a logical value, either TRUE or FALSE telling us if the value inputted is that type.

```
is.numeric(2)
# [1] TRUE
is.character("2")
# [1] TRUE
is.character(2)
# [1] FALSE
is.logical(TRUE)
# [1] TRUE
```

So far we've just been checking the value of a single thing: a single number, a single character/string, or a single logical/Boolean value. In practice almost everything we do will be on a column of a data set. These functions still work in the exact same way. We input the column (using the data$column syntax discussed in Chapter 2 to specify which data set we want and specifically which column in that data set) and the function will behave just like it did above. That's because each column can only be a single type of data; if the column is numeric, all values will be numeric; if the column is character, all values in that column are character; if the column is logical, every value in that column is also logical.

Let's use the UCR data from 2017 that was introduced in Chapter 2. Remember that the data must be in your working directory to load it. And here I have "data/" before the data name because the data is in a folder called "data" in my working directory. For more on working directories, please see Section 2.1.2.

```
load("data/ucr2017.rda")
```

We need to know the column names before using them, so we can use the `names()` function to get a list of all of the column names (the `colnames()` function does the same thing).

```
names(ucr2017)
# [1] "ori"                        "year"
# [3] "agency_name"                "state"
# [5] "population"                 "actual_murder"
# [7] "actual_rape_total"          "actual_robbery_total"
# [9] "actual_assault_aggravated"
```

Now we can check the types of some of the columns. Let's check the year column as an example. A year is a number so we may expect it to be numeric, but there's technically nothing stopping that data from being character type. It can't be logical type because then instead of a year value it'd just be TRUE or FALSE, which is certainly not what a year is.

```
is(ucr2017$year)
# [1] "numeric" "vector"
```

And we can use `is.numeric()` as another way to see if this column is numeric.

```
is.numeric(ucr2017$year)
# [1] TRUE
```

3.3   Data structures

We'll look in detail about two important data structures - vectors and
data.frames - and then talk briefly about two other structures that are not
that important in this book, but are nonetheless good to know that they exist. So far we've just been looking at either a single value, such as a <- 1 or
more complicated structures such as the ucr2017 data set, which is called a
data.frame - R's version of an Excel file. Data structures each operate a little
differently from each other so it's good to understand what they are and how
they work. We'll cover much more of how they work in Chapter 10, which covers how to subset data - which is just how to keep only certain values (such
as specific rows or columns) in the data.

3.3.1   Vectors (collections of "things")

The first data structure we'll discuss is a vector. A vector is a collection of
same type (numeric, character, logical, Date) values in a single object. When
we made "a" in Chapter 2, we assigned it only a single value, such as a <-
1. Usually we'll want to have a group of values - such as a set of years or a
group of crime types - rather than just a single value. We can do this by using
the same assignment method as a <- 1 but put all of the values we want to
assign to a into the function c() and separate each value by a comma. The c()
function combines each value together into a single vector.

Now, technically a single value, such as our object called "a" which now equals
1, is still a vector. In this case it'd be a vector of length 1, since there is only
one value in it. But when we generally talk about vectors there are multiple
elements in it.

Here's an example of making the object a be a vector with three values: 1, 2,
and 3 (in that order).

```
a <- c(1, 2, 3)
```

It is absolutely crucial to have the c() function, otherwise we'd get an error
from R.

```
a <- (1, 2, 3)
# Error: <text>:1:8: unexpected ','
# 1: a <- (1,
#                ^
```

It is likewise crucial to have a comma separating every single separate value.

```
a <- c(1 2 3)
# Error: <text>:1:10: unexpected numeric constant
# 1: a <- c(1 2
#                    ^
```

The terminology for talking about values in a vector is that each value is called an "element," and we identify them by the number they are, in order from start to finish. So here we have 1, 2, and 3, and we can say that the first element is 1, the second element is 2, the third element is 3.

If we assigned a to b (b <- a) we don't need to use the c() again. a is already a vector so if we assign its value to something, that carries over the vector. The c() is only necessary when first creating the vector.

Note that vectors take values that are the same type, so all values included must be the same type, such as a number or a string. If they aren't the same type, R will automatically convert it to the same type.

```
c("cat", "dog", 2)
# [1] "cat" "dog" "2"
```

Above we made a vector with the values "cat", "dog" and 2 (without quotes) and it added quotes to the 2. Since everything must be the same type, R automatically converted the 2 to a string of "2".

3.3.2 Data.frames

Nearly everything you do in this book and in research will be through data.frames. A data.frame is basically R's version of an Excel file. More precisely, a data.frame is a collection of equal-length vectors. Each column in a data.frame is actually a vector. They must all be equal length so every column has the same number of rows. You can't have, for example, a data.frame with 10 rows of data for the city column and only 8 rows for the year column. It must be 10 for each. Since vectors can only be a single type, each row in a particular column in a data.frame must be the same type, though different columns can be different types. This is how we can have, for example, our ucr2017 data.frame, which has both numeric and character type columns.

In this book I'll refer to data.frames by keeping it all lower case and with a dot between the words. This is just because the function to make one is data.frame(), and writing it this way is the normal convention. But writing it as a data frame is also fine. In nearly all cases we'll be using data that is loaded into R and is already in the structure of a data.frame (usually these will be Excel files or R data files like an .rda or .rds file).

If we wanted to create our own data.frame we would use the `data.frame()`, function and the input would be vectors, which will become our columns. Let's make a simple one. If the vector is already created then R would automatically take the name of that vector object as the column name, otherwise we could name it ourselves

```
example <- data.frame(
  column_1 = c(1, 3, 5, 7, 9),
  column2 = c(
    "hello",
    "darkness",
    "my",
    "old",
    "friend"
  )
)
example
#    column_1  column2
# 1         1    hello
# 2         3  darkness
# 3         5       my
# 4         7      old
# 5         9   friend
```

Now we have a new data.frame called example, which has two columns and five rows. We named the columns ourselves, and in this case we don't need to put the column name in quotes, though doing so would give the same result. Here we're saying that the column "column_1" is equal to the vector `c(1, 3, 5, 7, 9)` and "column_2" is equal to the vector `c("hello", "darkness", "my", "old", "friend")`. We're essentially creating an object inside of the `data.frame()` function but in this case we need to use the equal sign and not the `<-` because R doesn't allow the use of `<-` inside of a function.

If we forget to name the columns, and our vectors aren't already created with their own name, R will create a name based on the values in that vector. As shown below, this looks really bad so make sure to always name your columns.

```
example <- data.frame(
  c(1, 3, 5, 7, 9),
  c(
```

```
    "hello",
    "darkness",
    "my",
    "old",
    "friend"
  )
)
example
#    c.1..3..5..7..9.
# 1              1
# 2              3
# 3              5
# 4              7
# 5              9
#    c..hello....darkness....my....old....friend..
# 1                                        hello
# 2                                     darkness
# 3                                           my
# 4                                          old
# 5                                       friend
```

If the vectors are already made then we won't have an issue. R will default to the vector name, but we can override that if we want.

```
column_1 <- c(1, 3, 5, 7, 9)
column2 <- c("hello", "darkness", "my", "old", "friend")
example <- data.frame(column_1,
  overridden_name = column2
)
example
#    column_1 overridden_name
# 1         1           hello
# 2         3        darkness
# 3         5              my
# 4         7             old
# 5         9          friend
```

As with other objects, we can use the `is()` function to see what type it is. If we use `is()` on our example object, it'll tell us that it is a data.frame.

```
is(example)
# [1] "data.frame" "list"      "oldClass"  "vector"
```

We also often will want to know how many columns and rows a data.frame has. For finding the number of rows we use the function `nrow()`, and for finding the number of columns we'll use the `ncol()` function.[1] In each the "n" part of the function just stands for number. So `nrow()` is number of rows. For each we put our data.frame object in the parentheses (without quotes since it is something already loaded in R), and it will return the number of rows/columns.

```
nrow(example)
# [1] 5
```

```
ncol(example)
# [1] 2
```

Alternatively, we could have looked in the Environment tab which shows us the number of rows and columns of each data.frame that is loaded to R. For example, ucr2017 says it has "15764 obs. of 9 variables". This just means there are 15,764 rows and 9 variables. A variable in this context is just another way to say a column. However, you'll occasionally want to find the exact number of rows and columns, and as you'll often delete certain rows and columns from your data this can change throughout your code. So being able to use `nrow()` and `ncol()` is easier than repeatedly checking the Environment tab.

You may encounter something called a data.table or a tibble. These are two popular variations of data.frames that operate much the same way as data.frames but with some different features. We'll use tibbles in this book so will discuss their features when we use them.

3.3.3   Other data structures

There are two other data structures that I'll mention only so you have heard of them and can look up more information on them if you'd like. However, these are not *that* important to know about for the purpose of this book. Some of

---

[1]We could also use the `dim()`, function which tells the dimensions of the data.frame. The dimensions are the rows and columns in the data.frame so `dim()` tell us the results of both `nrow()` and `ncol()` at the same time. This function returns a vector showing first the number of rows and then the number of columns. But I find it easier to simply ask for the number of rows or columns separately, and to not deal with the result, which has two values.

these structures may come up in rare cases when you're programming, so it's important to know that they exist.

The first data structure is a list. A list is essentially a vector but where different values can be different types. Lists are actually very powerful data structures and ones that you'll encounter a lot when using R, but are almost entirely on the backend of R so not things you'll actually deal with much. For example, all data.frames are actually lists. And more specifically, they are a list of vectors. Lists can come in handy because they can store different types of data structures. A single list can, for example, have a number, a vector, a matrix (discussed below), and an entire data.frame inside. Lists can even have other lists inside of them. Let's look at an example of this.

```
list_example <- list(
  "hello",
  1:5,
  6:10,
  list(c(33, 66, 99)),
  head(mtcars)
)
head(list_example)
# [[1]]
# [1] "hello"
#
# [[2]]
# [1] 1 2 3 4 5
#
# [[3]]
# [1]  6  7  8  9 10
#
# [[4]]
# [[4]][[1]]
# [1] 33 66 99
#
#
# [[5]]
#                      mpg cyl disp  hp drat    wt  qsec vs am
# Mazda RX4           21.0   6  160 110 3.90 2.620 16.46  0  1
# Mazda RX4 Wag       21.0   6  160 110 3.90 2.875 17.02  0  1
# Datsun 710          22.8   4  108  93 3.85 2.320 18.61  1  1
# Hornet 4 Drive      21.4   6  258 110 3.08 3.215 19.44  1  0
# Hornet Sportabout   18.7   8  360 175 3.15 3.440 17.02  0  0
# Valiant             18.1   6  225 105 2.76 3.460 20.22  1  0
#                     gear carb
```

```
# Mazda RX4              4    4
# Mazda RX4 Wag          4    4
# Datsun 710             4    1
# Hornet 4 Drive         3    1
# Hornet Sportabout      3    2
# Valiant                3    1
```

The list that I called list_example contains six different elements in it: a character, two numeric vectors, a list of a numeric vector, and the first six rows of the mtcars data.frame. Lists can be useful when storing many different objects at once, but as they are not used too often for research-related programming I'll say no more of them.

The other type of data structure is a matrix. A matrix is a two-dimensional object where every value is the same type. Think of a data.frame but each column has to be the same type. Below is an example of a matrix with values 1 through 50 and with five columns and five rows. Every value here is a number.

```
matrix(1:50, nrow = 5, ncol = 5)
# Warning in matrix(1:50, nrow = 5, ncol = 5): data length
# differs from size of matrix: [50 != 5 x 5]
#        [,1] [,2] [,3] [,4] [,5]
# [1,]    1    6   11   16   21
# [2,]    2    7   12   17   22
# [3,]    3    8   13   18   23
# [4,]    4    9   14   19   24
# [5,]    5   10   15   20   25
```

If I change it to have the first value be "1" (in quotes so it is a character) and the others be the numbers 2 through 50, the matrix will automatically convert everything to a character type. So it will remain having everything be the same type, but now everything is a character.

```
matrix(c("1", 2:50), nrow = 5, ncol = 5)
# Warning in matrix(c("1", 2:50), nrow = 5, ncol = 5): data
# length differs from size of matrix: [50 != 5 x 5]
#        [,1] [,2] [,3] [,4] [,5]
# [1,] "1"  "6"  "11" "16" "21"
# [2,] "2"  "7"  "12" "17" "22"
# [3,] "3"  "8"  "13" "18" "23"
# [4,] "4"  "9"  "14" "19" "24"
# [5,] "5"  "10" "15" "20" "25"
```

4

Reading and writing data

For this chapter you'll need the following files, which are available for download here[1]: fatal-police-shootings-data.csv, fatal-police-shootings-data.dta, fatal-police-shootings-data.sas, fatal-police-shootings-data.sav, sqf-2019.xlsx, sf_neighborhoods_suicide.rda, and shr_1976_2020.rds.

So far in these lessons we've used data from a number of sources, but which came as .rda or .rds files, which are the standard R data formats. Many data sets, particularly older government data, will not come as .rda or .rds files but rather as Excel, Stata, SAS, SPSS, or fixed-width ASCII files. In this brief lesson, we'll cover how to read these formats into R as well as how to save data into these formats. Since many criminologists do not use R, it is important to be able to save the data in the language they use to be able to collaborate with them.

In this lesson we'll load and save multiple files into R as examples of how R can handle data that is used in many different software programs.

When loading data into R remember that your data must be in your current working directory or R won't be able to read it. For a refresher on working directories please see Section 2.1.2. In these examples I have my data in a folder called "data" that is in my working directory, which is why I use "data/" when naming the file. You do not need to include "data/" when loading in data on your computer.

4.1 Reading data into R

4.1.1 R

4.1.1.1 .rda and .rdata files

As we've seen earlier, to read in data with a .rda or .rdata extension you use the function `load()` with the file name (including the extension) in quotation

---

[1]https://github.com/jacobkap/crimebythenumbers/tree/master/data

marks inside of the parentheses. This loads the data into R and calls the object the name it was when it was saved. Therefore we do not need to give it a name ourselves.

Below, we're loading the "sf_neighborhoods_suicide.rda" file, and it creates an object in R (which we can look at in the Environment tab) called "sf_neighborhoods_suicide". It has the same name only because when I originally saved the file I saved it using the same name as it was called in R. But in practice I could have called it whatever I wanted. So it being the same name is convenient, as it is clear what the data is, but not necessary.

```
load("data/sf_neighborhoods_suicide.rda")
```

4.1.1.2   .rds files

For each of the other types of data we'll need to assign a name to the data we're reading in. Whereas we've done x <- 2 to say x gets the value of 2, now we'd do x <- DATA where DATA is the way to load in the data, and x will get the entire data set that is read in.

This includes the other kind of R data file, the .rds file. Here, we must explicitly name the data - there is no name by default like in a .rda or a .rdata file. We can load .rds files into R using the readRDS(), which is built into R so we don't need any package to use it. Like in load(), we just put the name of the file (in quotes) in the parentheses. Here we're naming it "rds_example," but we can name it whatever we like.

```
rds_example <- readRDS("data/shr_1976_2020.rds")
```

4.1.2   Excel

To read in Excel files that end in .csv, we can use the function read_csv() from the package readr (the function read.csv() is included in R by default so it doesn't require any packages but is far slower than read_csv() so we will not use it).

```
install.packages("readr")
```

```
library(readr)
```

The input in the () is the file name ending in ".csv". As it is telling R to read a file that is stored on your computer, the whole name must be in quotes. Unlike loading an .rda file using `load()`, there is no name for the object that gets read in so we must assign the data a name. We can use the name *shootings* as it's relatively descriptive for what this data is and it is easy for us to write.

```
shootings <- read_csv("data/fatal-police-shootings-data.csv")
```

`read_csv()` also reads in data to an object called a `tibble`, which is very similar to a data.frame but has some differences in displaying the data. If we run `head()` on the data it doesn't show all columns. This is useful to avoid accidentally printing out a massive amounts of columns.

```
head(shootings)
# # A tibble: 6 x 14
#        id name        date      manne~1 armed   age gender race
#     <dbl> <chr>       <date>    <chr>   <chr> <dbl> <chr>  <chr>
# 1      3 Tim Ell~ 2015-01-02 shot    gun      53 M      A
# 2      4 Lewis L~ 2015-01-02 shot    gun      47 M      W
# 3      5 John Pa~ 2015-01-03 shot a~ unar~    23 M      H
# 4      8 Matthew~ 2015-01-04 shot    toy ~    32 M      W
# 5      9 Michael~ 2015-01-04 shot    nail~    39 M      H
# 6     11 Kenneth~ 2015-01-04 shot    gun      18 M      W
# # ... with 6 more variables: city <chr>, state <chr>,
# #   signs_of_mental_illness <lgl>, threat_level <chr>,
# #   flee <chr>, body_camera <lgl>, and abbreviated variable
# #   name 1: manner_of_death
```

We can convert it to a data.frame using the function `as.data.frame()` though that isn't strictly necessary since tibbles and data.frames operate so similarly.

```
shootings <- as.data.frame(shootings)
```

To read in Excel files that end in .xls or .xlsx, we need to use the `readxl` package and use the `read_excel()` function. We'll read in data on stop, question, and frisks in New York City.

```
install.packages("readxl")
```

```
library(readxl)
```

```
sqf <- read_excel("data/sqf-2019.xlsx")
```

4.1.3 Stata

For the next three files, we'll use the package haven.

```
install.packages("haven")
```

```
library(haven)
```

haven follows the same syntax for each data type and is the same as with read_csv() - for each data type we simply include the file name (in quotes, with the extension) and designate a name to be assigned the data.

Like with read_csv(), the functions to read data through haven all start with read_ and end with the extension you're reading in.

- read_dta() - Stata file, extension ".dta"
- read_sas() - SAS file, extension ".sas"
- read_sav() - SPSS file, extension ".sav"

To read the data as a .dta format we can copy the code above that read in the .csv file but change .csv to .dta and change the function from read_csv() to read_dta().

```
shootings <- read_dta("data/fatal-police-shootings-data.dta")
```

Since we called this new data *shootings*, R overwrote that object (without warning us!). This is useful because we often want to subset or aggregate data and call it by the same name to avoid making too many objects to keep track of, but watch out for accidentally overwriting an object without noticing!

4.1.4 SAS

```
shootings <- read_sas("data/fatal-police-shootings-data.sas")
```

4.1.5 SPSS

```
shootings <- read_sav("data/fatal-police-shootings-data.sav")
```

4.1.6 Fixed-width ASCII

The final type of data source we'll talk about is a fixed-width ASCII. An ASCII file is just a text file and the fixed-width part means that each row has the exact same number of characters. This is a very old file format system that hopefully you'll never encounter but is one that some government agencies - including the FBI for their annual data releases (though some individuals and organizations re-release the data in better formats like R and Stata files) - still use, so it is good to know to how handle. A fixed-width ASCII file is essentially an Excel file but with all of the columns smushed together. It also tries to reduce its file size by replacing long strings of text with short ones. For example, instead of including a state name it'll usually have a number - as a number has fewer characters than an entire name, the file is therefore smaller.

Each fixed-width ASCII file also comes with what is called a "setup" file, which is some code that tells the program you're reading the data into how to separate columns and when to replace the (in our example) numbers that indicate the state with the actual state name. In nearly all cases where you have a fixed-width ASCII, you'll also be able to download the setup file from the same source, so I won't cover how to make a setup file yourself.

To read fixed-width ASCII files into R we'll use the `asciiSetupReader` package, which I created myself for this very purpose. For more information on this package including details on all of the different options in the function, please see the package's site here.[2]

---

[2]https://jacobkap.github.io/asciiSetupReader/

```
install.packages("asciiSetupReader")
```

We'll use the `read_ascii_setup()` function, which takes two mandatory inputs in the parentheses: the name of the data file (which will have a file name ending in .txt or .dat) and the name of the setup file (which will have a file name ending is .sps or .sas). Each of these file names must be in your current working directory, and you must put the names in quotes. The data file in this example is the 2020 FBI Supplementary Homicide Report data (their murder data set), which is called "2020_SHR_NATIONAL_MASTER_FILE.txt" and the setup file is called "ucr_shr.sps". We can name the object we read "shr".

```
library(asciiSetupReader)
shr <- read_ascii_setup(
  "data/2020_SHR_NATIONAL_MASTER_FILE.txt",
  "data/ucr_shr.sps"
)
```

4.2   Writing data

When we're done with a project (or an important part of a project) or when we need to send data to someone, we need to save the data we've worked on in a suitable format. For each format we are saving the data in, we will follow the same syntax of

```
function_name(data, "file_name")
```

As usual we start with the function name. Then inside the parentheses we have the name of the object we are saving (as it refers to an object in R, we do not use quotations) and then the file name, in quotes, ending with the extension you want.

For saving an .rda or .rdata file we use the `save()` function. For saving a .rds file we use the `saveRDS()` function. Otherwise we follow the syntax of `write_` ending with the file extension.

- `write_csv()` - Excel file, extension ".csv"
- `write_dta()` - Stata file, extension ".dta"
- `write_sas()` - SAS file, extension ".sas"
- `write_sav()` - SPSS file, extension ".sav"

As with reading the data, `write_csv()` comes from the `readr` package while the other formats are from the `haven` package. Though the `readxl` package lets you read .xls and .xlsx files, it does not currently have functions that let you save a file to that type.

There are other packages that let you save .xls and .xlsx file but in the interest of keeping the packages we learn to a minimum, I won't include those here. In nearly all cases you'll want to save your data as an .rds, .csv, or a .dta file. Fixed-width ASCII files are so primitive that while we may need to load them into R, we should never save data in this format.

4.2.1   R

4.2.1.1   .rda and .rdata

For saving an .rda file we must set the parameter `file` to be the name we're saving. For the other types of data they use the parameter `path` rather than `file` but it is not necessary to call them explicitly. A parameter in a function is just an option for how the function works. Only for `save()` do we need to write `file =` explicitly in the function.

```
save(shootings, file = "data/shootings.rda")
```

4.2.1.2   .rds

```
saveRDS(shootings, "data/shootings.rds")
```

4.2.2   Excel

```
write_csv(shootings, "data/shootings.csv")
```

4.2.3   Stata

```
write_dta(shootings, "data/shootings.dta")
```

4.2.4   SAS

```
write_sas(shootings, "data/shootings.sas")
```

4.2.5   SPSS

```
write_sav(shootings, "data/shootings.sav")
```

Part II

Project Management

5

Mise en place

If you're familiar with cooking you might have heard the phrase *mise en place*,[1] which is French for "everything in its place." In cooking this concept means that you get everything - ingredients, pots, pans, bowls, utensils, etc. - needed to cook that item ready before you begin cooking. This saves time as you have everything you need in front of you and can just cook from start to finish without stopping to find something. This is also a useful idea in programming, especially when you're programming to conduct research.

In this chapter, we'll cover how to get *mise en place* for your programming projects. First, we'll discuss how to think about the project and write out each step that we need to take to complete the project, and each output that we want (such as a graph or table). We'll write this out by hand and in plain English (or whichever language you are most comfortable in), before writing any code. Finally, we'll go over what is, in my opinion, the best way to organize your folders, data, and code. This method is particularly suited for research projects, but please feel free to modify my methods to suit your own needs and preferences.

5.1   Starting with a pencil and paper

This may seem counter intuitive, but the best way to start any programming project - and in particular, research project - is to use a pencil and paper. On this paper you should outline every step (broadly speaking, not literally every line of code) that you'll take for the project. This is a useful process at the start of a project to step back from the code and think about the overarching goal of the project - and what you need to do to get there. For example, let's think about doing research using data from the US Border Patrol data (We'll actually work on this data in Chapter 22). The US Border Patrol releases data as PDFs, which have a table showing the annual number of apprehensions they make. We want to see if a policy change affected apprehensions at the border. On the data side, that'd require scraping and cleaning the PDFs.

---

[1] https://en.wikipedia.org/wiki/Mise_en_place

On the analysis side, we'd probably want to do a time-series graph showing apprehensions over time, and run a regression to see if the policy had a significant effect. So here we have four broad categories of work (scraping, cleaning, graphing, running a regression) for a fairly simple policy evaluation. Within each category you can make a number of subcategories of steps you need to do. For example, in the scraping category you might want to add the following subcategories: download the PDFs, see how each table relates to each other, figure out which parts of the tables are actually relevant, etc. We can probably break down these subcategories even further if we want.

You essentially want to build a roadmap to follow - you can, of course, deviate from this roadmap if necessary - as you work on the project. This is useful for two reasons. First, writing out what you need to do will often clarify exactly what you need to do. Knowing that you'll want a time-series graph, for example, will mean that you need to have your data aggregated into a certain time unit. Knowing this before-hand will save you time as you'll have a tangible goal to work towards and don't have to keep stopping during your work to figure out what to do next.

And second, from my experience helping people at Penn with R, people - especially new programmers (and myself when I first started learning R, my first programming language) - can get overwhelmed with programming. One major problem they had is they couldn't articulate what they needed to do since they weren't familiar enough with R to know the right words.[2] They knew the end goal, and what they had at the start, but couldn't articulate the path from start to finish. Writing out each step in plain language allowed them to know the path - it is simpler to know what steps you need to do to complete a project in plain language than to actually write the code (though this still requires experience to tell you a lot of the "minor" intermediate steps). Having a game plan helps people avoid being overwhelmed since they could do one step at a time (and feel accomplished at each step).

5.1.1 Tables and graphs

One of the biggest challenges I had early in my PhD was figuring out what data was supposed to look like. I mean that literally. My first research project was analyzing if monthly crime in school buildings changed after a new policy was instituted that increased building security. The data I had available was incident-level so one row for every crime at the school, and I needed to convert it to the building-month level. For some reason I just couldn't think of the

---

[2]Knowing the "right words" is surprisingly important when it comes to programming because it is crucial to finding help online. If you don't know how to subset, you can easily find how to subset in R on Google. But if you don't know the word subset or how to describe what subsetting does well, it can be very tricky to find help (you won't even know what to Google!). These "right words" are, annoyingly, an important part of programming that I believe isn't given enough focus.

proper way for my data to appear in the final data set, which prevented me from figuring out what I needed to do. One solution to this - and useful even if you don't have this problem - is to draw out the graphs and tables you want before starting the code. Like writing out the steps for the code, drawing the graph will help you understand exactly how your data needs to look - and thus what code you need to write - for these graphs.

Below are two images from a recent project of mine with the tables and graphs that I wanted sketched out. Note that in the image showing my graphs I have crossed out the first graph. These sketches are just preliminary tools to help your work, you aren't chained to them. Like any tool, if it is no longer relevant or useful, find something new. For regression result tables especially, sketching these out helps you think about what variables you will need to have to run the regression. For example, you may want to have control variables for demographics in your geographic unit (say, from the US Census). If we continue our example of using the US Border Patrol data, this means that you'll also need to grab, clean, and merge Census data to your other data sets. Sketching out the resulting tables and graphs is a good tool to figure out steps that you'll need to do for the project but may have not thought of.

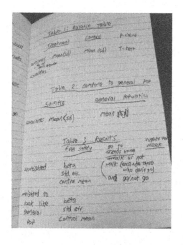
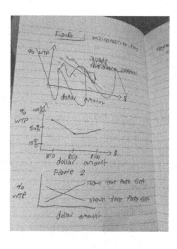

5.2   R Projects

We've talked about projects in an abstract sense - that they are research papers or specific data exploration jobs. RStudio provides, a bit confusingly, something called an R Project, which is merely a helpful way to organize folders for a specific project (paper, data exploration, etc.) that you do. When you do a project, I recommend keeping *everything* for that project in a single folder on your computer. Below is an image showing all of the folders I use for my various R work. As you can see from the file names, each folder is for a separate project, and there is not overlap between them - each project is independent.

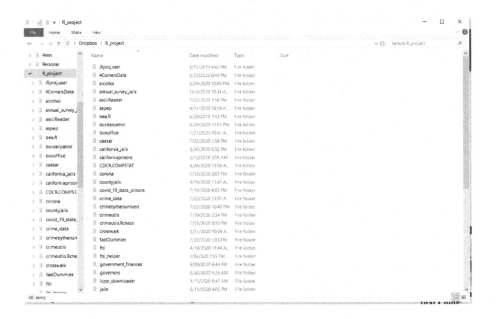

First, I'll explain how to set up an R Project through RStudio, and why you would want to do it. There are two main reasons to want to use an R Project. First, throughout this book I had you set your working directory so that R knew where to look for a particular file. In R Projects, by default the working directory is in that project's folder. So if you had a file example.csv in your project folder, you wouldn't need to set a working directory since R would already be looking in that folder.

This may be a minor time-saving method if you're working alone since you'd only need to set the working directory once when not using an R Project. But consider if you're collaborating with three people and you've shared your code. When using an R Project, it just runs. Your collaborators won't need to change the working directory to their own directory. Second, it provides easy access to using the version control software Git, which we'll talk about in detail in Chapter 9.

To make an R Project, start by clicking the *File* button on the top-left corner of RStudio and then click *New Project*. This will open up a window that has three options: New Directory, Existing Directory, and Version Control. New Directory says that the project we are making is going to be in a brand-new folder that we're (R will do this automatically) going to create. This is the one you'll click on in the majority of cases. Existing Directory is for making a folder in an existing folder, which doesn't have too many useful cases. The Version Control is taking a project that someone else has created and downloading it to your computer. We'll cover this more in Chapter 9.

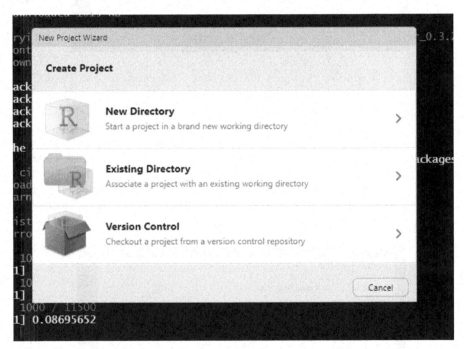

Once you've clicked New Directory, it'll change the window to ask you what type of project you want. The following two figures show all the different types of projects R can make (installing some R packages, such as bookdown, can add more types of projects to this list). R is very versatile and has project types ranging from the standard R Project to books and websites. We just want a standard project so click the New Project button at the top.

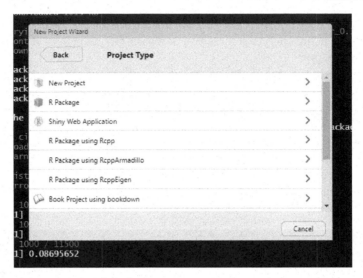

Now it'll have a window that says Create New Project up top. In the Directory name: section you write the name of your R Project. This will be the name of your folder so you want it descriptive enough to understand (and for collaborators to understand) what it is for, without being overly long. Once you have a name you can click the Browse... button on the right and go to the folder on your computer where you want to put this folder (ideally, you'll put it in a folder that is backed up by something like DropBox). Make sure the *Create a git repository* checkbox is selected, and we'll explain why in Chapter 9. Click Create Project and R will make the project folder on your computer and open that project in RStudio.

Below are images of a brand new R Project that I made called *example* that I put in my Desktop folder. The folder is now empty except for two files - .gitignore (which we won't talk about here) and *example* which is type "R Project" (and the full name would be *example.Rproj*). This is a **very** important file. Note that its name is the same as the R Project name that I made, and the same as the folder name on my computer. This file is essentially a shortcut that you click to open that R Project. It doesn't do anything more than open the R Project, but this is the way you'll access the project every time you want to use it. Double-click this and the R Project will open.

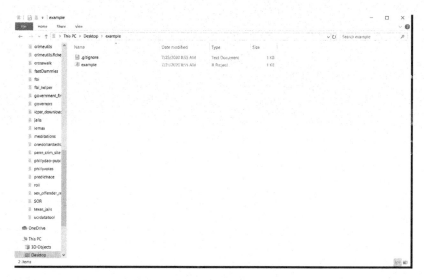

This RStudio session looks nearly identical to other sessions that we've used - and it is nearly identical. A few key differences can be found in the top-left corner where it says "example - RStudio," indicating that we're in the *example* R Project. And then directly below the "Console" tab it says "C:/Users/user/Desktop/example/". This is the working directory of this project. I didn't set it; R just knew where it was. If you move this folder to a new folder (say, the Downloads folder) or if someone else downloads it to their computer, R will automatically change the working directory to the right one. You no longer have to worry about it.

5.2.1 Folders

Now that we have the R Project made, we need to start adding some R code and data files to the project so we can get started working. But first, let's talk about proper ways to organize the folder. I've added a few new folders to the new *example* R Project as the basic layout of my work process. This is for a research-oriented project so it may not apply in your particular case. Organizing your folders (and as we'll see below, your code) is important so please play around with different ways to organize and find a way that works well for you.

I've added five folders to the R Project folder: analysis, articles, data, drafts, and R (note that I moved it to the Downloads folder, and if I opened the project RStudio would know where the new working directory was). I tend to do my analysis using Stata (primarily because most of my co-authors use Stata instead of R so this is a way we can both work on the analysis) so in the analysis folder I'd keep all of the .do (Stata) files to run the regressions. In articles, I put PDFs of every article I read that I use (or planned to use while reading it) for the paper I'm working on in this project. It's good to keep this organized to share with co-authors or just for easy reference after you've read it. It certainly takes time to find good sources for a lot of research, so you don't want to have to search again because you've forgotten which article you had a particular reference from or that was important to your study. While I recommend writing your papers in R Markdown (see Chapter 7), you will need to create drafts of the paper to send to others (e.g. your collaborators or journals). The "drafts" folder is a good place to keep these versions - some journals require that you submit a Word Document with track-changes for a revise and resubmit so you will need to leave R Markdown occasionally to comply with these rules.

The final two important folders are "R" and "data." In the R folder - as you may have guessed - belong the various R scripts that you write during the project. In Section 5.3, we'll talk in detail as to how to organize these scripts. Inside the "data" folder I made two subfolders: "raw" and "clean." The raw folder is where you'll store the data exactly as you got it (for cases where the data is acquired through webscraping, this isn't necessary). This folder will have, for example, the PDFs that you intend to scrape, or .csv files with crime data in them. It is important to keep this data always unchanged (change it only in R and save the output to a new file) so you can replicate your results from the original data. In the clean folder is that final data output from your work to clean and manipulate (e.g. subset, aggregate) the data. It isn't strictly necessary to even output a final data set - you could just rerun your code from the original data each time, and this is fine if your code is very quick to run - but it is important both for safekeeping and to be able to share with others. If you collaborate with people, you'll want to be able to send them the data so they can examine it without having to run all of your code themselves.

· This PC > Downloads > example > data

Name	Date modified	Type	Size
clean	7/23/2020 9:04 AM	File folder	
raw	7/23/2020 9:04 AM	File folder	

5.3   Modular R scripts

If you are like many people who start programming, all of your code will be in a single R script. This is fine when you're first getting familiar with R and don't want to go searching for code in places when you're still uncomfortable with the language. As you become more familiar with R - and as your projects get more complex - you'll want to start making multiple R scripts in a single project.

When you're writing a paper you don't just write one extremely long sentence. You break up ideas into paragraphs and divide groups of paragraphs into larger sections. This is useful in a paper to organize your thoughts and to make it readable for others. It's also useful when working since you know, for example, "Section 1 is done, but I still need to finish Section 2 and the last part of Section 3." This way you don't confront working on the entire paper at once. You'll want to follow these lessons in the code you write, with each "section" of code being its own R script and within a script split up code into particular "paragraphs." The end goal should be to have modular R scripts, with each script being independent (or relatively so) and the combination of these parts has all the code for your particular project. This is a bit of an abstract concept so let's use a real example from one of my recent projects.

Above is a folder for the code used to analyze data for a paper examining perceptions of outdoor lighting. There are five R scripts in the folder - clean.R, census.R, tables.R, graphs.R, and utils.R - and these are the only ones used for this project.[3] Each of these files (utils.R is an exception) has a particular role to play in the analysis of the data. The first file, clean.R is just code that cleans up the survey data and makes it ready to be analyzed and graphed. The census.R file has code that cleans Census data that my co-author and I use to compare our survey sample to the general public. As this is a separate data set than the survey data, I have it in its own R script. tables.R and graphs.R are the code to make descriptive statistics tables and figures for the paper, respectively. I chose these files because they are doing fairly separate tasks, all with the goal of turning raw data into a research paper.

This is an example of how I approach making R scripts, not necessarily the best way to do so. Even here, other decisions could be made. For example, I could have put the code from census.R into clean.R since they're both about cleaning data. While you should try to make separate R scripts for broadly different tasks (regardless of how much code that task requires), you should experiment with how you prefer to separate these scripts, and balance between having one (or a few) super scripts that comprise everything with having too many scripts that do too little - this balance requires experience and experimentation so keep at it!

---

[3]The analysis was done in Stata so there are separate files for that.

5.4   Modular code

In addition to having separate scripts for each major part of your project, you will want to organize each individual script into relatively modular parts. Whereas each script is like a book chapter, the code inside the script should be like paragraphs, separated into distinct chunks. For example, let's say you have some raw data and want to subset it, change some values (e.g. renaming F to Female, M to Male), and then aggregate to a larger geographic area. This is a three-step process - subset, change, aggregate - so you'll want to have three different parts of your R script dedicated to this. Now, if this is a simple process (and it will always depend on the data and what you want to do with it), you may want to have each step in its own Section (as we'll discuss next). If it's relatively simple and takes only a few lines per step, you'll likely just want to have a line break between steps and identify your choices in comments. It's hard to give precise rules on how to do this as it really does depend on personal preference - I think having more comments and line breaks early in your R career is best as you're still learning R, and it is good practice to comment your code. You can always alter this balance to suit your preferences as you gain more experience.

The goal of making modular code is to avoid having a large amount of code without breaks or comments - that'd be like reading a run-on sentence. We'll talk about comments more in Section 6.2.1, but here you should explain your choices (e.g. "Subset to only violent crime and property crime") to inform collaborators (other people and yourself in the future who will likely forget what or why you did something), but without writing too much. Generally the rule of thumb is to have comments for why you did something, not explaining what you did. I think a mix of what and why is helpful as it's quicker than looking at the code, especially if your code is complex. Like a lot of your work, however, this depends on the project and your audience - if you're working with someone new to R, having more comments explaining what you did is helpful.

5.4.1   Section labels

When you have major parts of an R script, you should have something to indicate that this is a distinct section from other parts. RStudio has a handy tool to help make that distinction by creating sections in your R script. Press the keys Control+Shift+R (Command+Shift+R in a Mac), and it will open up a window where you can set a section label. Write the name of the section you want and click OK, and it'll add that to where your cursor was in the

R Script. You can also do this by simply adding four dashes on the end of a comment.

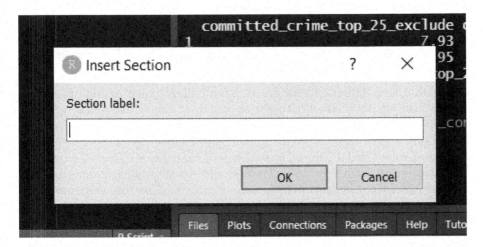

Sections are more than just commented parts of a Script. Note that in the following screenshot, there is both the Section label in the R Script and that same label in a new section of the Source tab on the right. You can get to this section by clicking on the button on the very top right, the one that looks like a bunch of misaligned lines. In here, it shows all the Sections that exist and clicking the Section name will move to the start of that Section in your R Script. If you have a long script (which is generally unadvised but sometimes can't be helped), this is an easy way to find a particular part of your code.

5.4.2   Helper R scripts

As part of making code organized, I find it helpful to make an R script in each (or most) of my projects to hold helper functions or objects - and I call this utils.R (utils stands for utility as these are helpful pieces of code for the project). This file should be for code that will be used in multiple R Scripts, so you want them in a single place rather than copying them over in each script where you need them.

In utils.R, I keep functions that are either auxiliary (such as code to check data by printing out a set of outputs) or code that is used infrequently (such as loading several files and merging them together at the start of an R script) where I don't want them in the main file. I also include useful objects such as a vector of values that I will use to subset. For example, if I wanted to subset all violent crimes from a data set, I would need to know what crimes in that data are considered violent, put them as strings in a vector, and subset to only rows that match those strings. I could make an object with this vector, such as `violent_crimes <- c("murder", "rape", "robbery", "assault")`.

If you want to run utils.R (or any .R file) in a different R script, you can use the `source()` function, which makes R run the entire script inputted in the parentheses. Just put the file name (in quotes) in the parentheses and it will run. For example, if we want to run utils.R, we'd write `source("utils.R")`. If that file was in the "data" folder of our R Project, we'd write `source("data/utils.R")` so R knew to look in the "data" folder for the file.

It isn't necessary to make helper functions like these, but I find them helpful. I recommend that you try them out when you do R projects, but if you don't find them useful please feel free to stop using them.

6

Collaboration

6.1 Code review

When you collaborate with other people, you will probably each be working on a separate (though related) part of the project and then will combine each part when you are done. Combining your code could be through emailing each other R scripts - and having one person combine everything - or something more formalized such as using Git, which we discuss in Chapter 9. However you decide to do this, it is important to use a process to review your collaborator's code (and have them review yours) to check for mistakes.[1] This is a similar process to having a colleague read a paper draft before submitting it.

Code review is a useful technique for reducing the number of mistakes as it is a check on the work before using the code for real. Code review generally involves having one person who writes the code send it to another person who checks the code for any potential mistakes or issues. This check involves ensuring that the code meets the specified style (this is discussed further in Section 6.1.1) and that there are no bugs (which are errors in the code). For the person having their code reviewed, having comments explaining the what and why of the code (discussed more in Section 6.2.1) will help the reviewer quickly go through the code. The code should also be relatively short, comprised of a specific R script (or related scripts) and no more than a few hundred lines of code. This is because as code gets more interrelated and complex, it is harder for someone unfamiliar with the code to understand it and see any issues. That means that a reviewer for long code is more likely to miss issues and take longer to review. Reviewing shorter code, even if that means reviewing more often, is often far more efficient for both the reviewer and reviewee and catches more issues.

In cases where you have unit tests (which are discussed in Chapter 8) written for the code, these tests are an automated form of code review as they too check for mistakes. To save people's time, you should avoid sending the code for review until it passes all unit tests. However, if you're stuck and can't get certain tests to pass, working with someone else to solve the problem is often

---

[1]If your collaborator does not know R, they should read this book.

faster than doing so yourself because then you have an outside perspective who may see something that you missed.

For code review to be most efficient, I recommend developing some rules with your collaborators to specify how and when code review is done. For example, you should determine who reviews certain people's code (ideally with senior people reviewing junior people's code) and how often it is done. I believe that doing code reviews relatively frequently (i.e. after a working draft of some code is ready) is useful as you can catch issues early and not waste anyone's (especially the person writing the code) time. However, having hard time limits is probably ill-advised as sometimes writing certain code takes far longer than expected and reviewing an unfinished (and potentially far from finished) bunch of code is not efficient for anyone.

When someone is reading a draft of your research paper, they are generally looking for whether it is correct (i.e. your methods are right, the lit review is thorough, etc.) and how well it flows. Code review is the same. While the primary goal is finding errors, an important aspect is to ensure that it is readable (i.e. proper spacing, how names are written) and consistent across everyone's code in that group. More formally, ensuring that everyone's code is readable and consistent is having people follow a style guideline.

6.1.1   Style guidelines

An important part of reviewing people's code is ensuring that everyone is following the same style guidelines when it comes to writing code. Style guidelines are the grammar rules of writing code. They dictate (or encourage) certain style choices, such as whether object names are lowercase, whether they include punctuation, and even when to put long code on a new line. This is equivalent to making sure that people writing in plain language put punctuation and capitalization in the expected place. While you can read !SomEThiNg WrITen. LiKE thIs, it is easier to understand when it follows adopted and accepted rules.

The important thing here is to be consistent. Consistency makes code much easier to read and helps make code written by multiple people more interchangeable. This book follows the tidyverse style guide,[2] which is one that many R programmers follow, but the exact style you choose is relatively unimportant (choosing more common styles helps when your code may be used by people out of your organization). Feel free to adopt an already-made style guide, make any modifications to suit your preferences, or to create an entirely new one yourself. As long as people follow the same format, you'll be able to spend more time on the code, and less time trying to understand it.

---

[2] https://style.tidyverse.org/

6.2 Documentation

An important, though occasionally tedious, part of writing code is documenting your work. We'll talk about documentation in two ways, through comments, which focus on specific parts of code, and vignettes, which document the project more broadly.

6.2.1 Comments

In Section 2.1 we introduced comments, which are essentially notes about the code that you include in an R script (by starting a line with the pound key #) that isn't run. They are just "comments" to yourself or anyone else reading the code to explain what that code does and why it is there. As is often repeated in explaining the benefit of comments, the main collaborator you will have is yourself in the future.[3] You don't need to comment on every single line of code - and doing so would just make it hard to read - but you should comment on important things or chunks of code (i.e. several lines of code that all are for the same purpose). If you write a function, you'll want at least a brief comment explaining what it does and what the inputs and parameters do.

Writing comments is not as fun as writing code. Stopping to write a comment on something that seems obvious at the time (after-all, you figured out how to do something you wanted to do and likely were focusing on) interrupts the flow of writing code and slows down your work. And when you have looming deadlines and multiple projects that you're working on, spending the time writing good comments may seem like a bad use of time as the payoff is only in the future. However, the benefits far outweigh the cost. This is true for two reasons. First, when you're collaborating with others, it is much quicker to have text explaining the code than to walk through the code with them (or to have them try to figure it out themselves).[4] As you work with more people, comments become increasingly important. Writing good comments is also time-efficient when considering that in many cases when you do research you will have to return to the project in the future.

This is best shown when considering a research project that leads to a journal article. For many papers, even if you are fantastically productive and can work nonstop at it without forgetting any decisions, at a certain point you'll

---

[3]I recently worked on a follow-up paper to one I had done a year ago. For some reason, past me decided to name some functions based on the authors of a paper that created that particular method, and didn't leave comments explaining what the code did or why. Past me caused a lot of problems for current me. Please comment your code!

[4]This is one of the main reasons I wrote this book. After a few years of helping Penn students with the same questions, I decided to write out guides to those topics.

need to finish and submit it to a journal. Journal reviews can often take three to six months so at that point you'll likely have forgotten many of the (seemingly obvious) decisions you made in the course of the project.[5] Having comments explaining why you made a certain decision (such as including or excluding certain crime types from your analysis) can be a huge time-saver when addressing reviewer concerns - you will know why each decision was made and won't have to try to figure out the *why*. This is particularly important when you have to defend a decision in which there is no obvious choice and you want to know your thought process at the time you wrote the code and were immersed in the issues of the data. A lot of data decisions are reasonable at the time based on the quirks of the data but can appear to make no sense if you aren't familiar with the data - comments can remind you of the quirkiness and how you handled it.

6.2.2   Vignettes

Vignettes are essentially a document that explains how to do something with the code you have written. This is common when someone has written an R package and they want to explain in detail important functions from the package. You can think of chapters of this book as vignettes covering particular topics - PDF scraping, webscraping, regular expressions, etc. To make a vignette, you can make an R Markdown file (for more information on R Markdown please see Chapter 7) detailing that topic. Since the text you write is included in the document, these files are basically normal R scripts with extensive comments written in plain language. Often, these comments are more formal than what you'd write in an R script as they are written as complete sentences or paragraphs and walk through comprehensive ideas rather than focus on discrete chunks of code.

One increasingly prominent method of using R for research is to do everything in an R Markdown file. This allows you to explain your approach - including context on why you did something - and each step you took in plain language in the text of the R Markdown file while still including the code directly in the file. Whether you include the code in the output (e.g. a PDF or Word Document), or just the result of the code (e.g. a graph or table), depends on your audience and how far along you are in the project.

If this is for a presentation to update collaborators, for example, it is useful to include the code as they may notice an issue or give advice based on the code. Including code can also teach your audience something new (I've certainly learned a lot by watching people present using code I wasn't familiar with). If the document is for an audience unfamiliar with R (or programming more

---

[5]If you're like me and on your seventh rejection for a particular paper, three to six months may be optimistic.

generally), or where time to present is limited, you probably won't want to include code.

Whether you do your work in an R script or in an R Markdown file is up to you. If you intend to write up a report anyway, having everything written up in the R Markdown file as you write your code can save you time as you're merging the code and the writing process. However, this loses some nice features in R such as unit tests, which we will discuss in detail in Chapter 8. It also depends on how complex your project is. If you have code that is hundreds of lines long and spans multiple R scripts, putting it all into a single R Markdown file is unfeasible. In this case it'd be better to run the code in the R scripts and use the R Markdown file just to present results.

7

R Markdown

When conducting research your end product is usually a Word Document or a PDF which reports on the research you've done, often including several graphs or tables. In many cases people do the data work in R, producing the graphs or numbers for the table, and then write up the results in Word or LaTeX. While this is a good system, there are significant drawbacks, mainly that if you change the graph or table you need to change it in R **and** change it in the report. If you only do this rarely it isn't much of a problem. However, doing so many times can increase both the amount of work and the likelihood of an error occurring from forgetting to change something or changing it incorrectly. We can avoid this issue by using R Markdown, R's way of writing a document and incorporating R code within.

This chapter will only briefly introduce R Markdown, for a comprehensive guide please see this excellent book.[1] For a cheat sheet on R Markdown see here.[2]

What R Markdown does is let you type exactly as you would in Microsoft Word and insert the code to make the table or graph in the places you want it. If you change the code, the document will have the up-to-date result already, reducing your workload. There is some additional formatting you have to do when using R Markdown but it is minimal and is well-worth the return on the effort. This book, for example, was made entirely using R Markdown.

I include this chapter early in the book - and likely before you are really comfortable with using R - since some new R programmers do like to do all of their work using this method. In my experience this is relatively rare, but I still wanted to make the info available for those that do. For new programmers I recommend reading this chapter so you understand R Markdown, but still use normal R scripts when writing code - don't use R Markdown for everything. Focus on learning how to write good code before adding the complexity of writing full documents using R Markdown.

To open up an R Markdown file click File from the top menu, then New File, and then R Markdown...

---

[1] https://bookdown.org/yihui/RMarkdown/
[2] https://www.rstudio.com/wp-content/uploads/2015/02/RMarkdown-cheatsheet.pdf

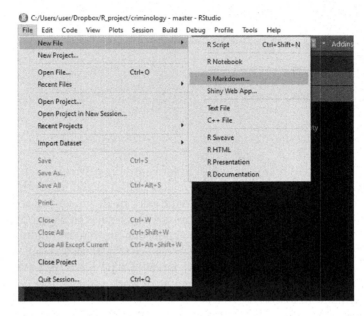

From here it'll open up a window where you select the title, author, and type of output. You can always change all three of these selections right in the R Markdown file after making your selection here. Selecting PDF may require you to download additional software to get it to output - some operating systems may already have the software installed. For a nice guide to making PDFs with R Markdown, see here.[3]

---

[3]https://medium.com/@sorenlind/create-pdf-.reports-using-r-r-markdown-latex-and-knitr-on-windows-10-952b0c48bfa9

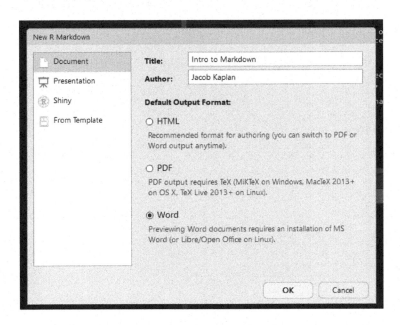

When you click OK, it will open a new R Markdown file that is already populated with example text and code. You can delete this entirely or modify it as needed.

When you output that file as a PDF it will look like the image below.

Intro to Markdown

Jacob Kaplan

March 11, 2018

R Markdown

This is an R Markdown document. Markdown is a simple formatting syntax for authoring HTML, PDF, and MS Word documents. For more details on using R Markdown see http://rmarkdown.rstudio.com.

When you click the **Knit** button a document will be generated that includes both content as well as the output of any embedded R code chunks within the document. You can embed an R code chunk like this:

```
summary(cars)
```

```
##      speed           dist
## Min.   : 4.0    Min.   :  2.00
## 1st Qu.:12.0    1st Qu.: 26.00
## Median :15.0    Median : 36.00
## Mean   :15.4    Mean   : 42.98
## 3rd Qu.:19.0    3rd Qu.: 56.00
## Max.   :25.0    Max.   :120.00
```

Including Plots

You can also embed plots, for example:

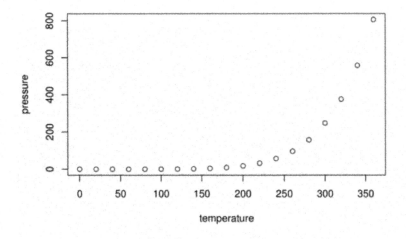

R converted the file into a PDF, running the code and using the formatting specified. In an R Script a # means that the line is a comment. In an R Markdown file, the # signifies that the line is a section header. There are 6 possible headers, made by combining the # together - a # is the largest header while ###### is the smallest header. As with comments, they must be at the beginning of a line.

The word "Knit" was surrounded by two asterisks * in the R Markdown file and became bold in the PDF because that is how R Markdown sets bolding - to make something italics using a single asterisks like *this*. If you're interested in more advanced formatting please see the book or cheat sheet linked earlier.

Other than the section headers, most of what you do in R Markdown is exactly the same as in Word. You can write text as you would normally and it will look exactly as you write it.

7.1 Code

The reason R Markdown is so useful is because you can include code output in the file. In R Markdown we write code in what is called a "code chunk". These are simply areas in the document which R knows it should evaluate as R code. You can see three of them in the example - at lines 8-9 setting a default for the code, lines 18-20 to run the summary() function on the *cars* data (a data set built into R), and lines 26-28 (and cut off in the screenshot) to make a plot of the data set *pressure* (another data set built into R).

To make a chunk click Insert near the top right, then R.

It will then make an empty code chunk where your cursor is.

Notice the three ' at the top and bottom of the chunk. Don't touch these! They tell R that anything in it is a code chunk (i.e. that R should run the code). Inside the squiggly brackets {} are instructions about how the code is outputted. Here you can specify, among other things if the code will be outputted or just the output itself, captions for tables or graphs, and formatting for the output. Include all of these options after the r in the squiggly brackets. Multiple options must be separated by a comma (just like options in normal R functions).

If you do not have the R Markdown file in the same folder as your data, you'll need to set the working directory in a chunk before reading the data (you do so exactly like you would in an R Script). However, once a working directory is set, or the data is read in, it applies for all following chunks. You will also need to run any packages (using library()) to use them in a chunk. It is good form to set your working directory, load any data, and load any packages you need in the first chunk to make it easier to keep track of what you're using.

7.1.1   Hiding code in the output

When you're making a report for a general audience you generally only want to show the output (e.g. a graph or table), not the code that you used. At early stages in writing the report or when you're collaborating with someone who wants to see your code, it is useful to include the code in the R Markdown output.

If you look at the second code chunk in the screenshot (lines 18-20) it includes the function `summary(cars)` as the code and the options `{r cars}` (the "cars" simply names the code chunk "cars" for if you want to reference the chunk - or its output if a table or graph - later, but does not change the code chunk's behavior). In the output it shows both the code it used and the output of the code. This is because by default a code chunk shows both. To set it to only show the output, we need to set the parameter `echo` to FALSE inside of the `{}`.

In the third code chunk (lines 26-28), that parameter is set to false as it is `{r pressure, echo=FALSE}`. In the output it only shows the graph, not the code that was used.

7.2   Inline Code

You can also include R code directly in the text of your document and it will return the output of that code. To use it, you need to setup an inline code chunk using the tick mark followed by the lowercase letter R, the code you want to use, and then end it using another tick mark. This is called using inline code. When you have a table or visualization to output, this isn't the proper method, it is best for small pieces of text to add to your document. This is most useful for when you want to include some descriptive info, such as the number of respondents to a survey or the mean of some variable, in the text of your document. Inline code will only present the output of the code and doesn't show the code itself. Below is an example of inline code - see the image below that for what it looks like with the code.

The data set mtcars has 32 rows and 11 columns. The mean of the mpg column is 20.090625.

```
The dataset mtcars has `r nrow(mtcars)` rows and `r ncol(mtcars)` columns. The mean of
the mpg column is `r mean(mtcars$mpg)`.
```

TABLE 7.1: This is an example table caption

	mpg	cyl	disp	hp	drat
Mazda RX4	21.0	6	160	110	3.90
Mazda RX4 Wag	21.0	6	160	110	3.90
Datsun 710	22.8	4	108	93	3.85
Hornet 4 Drive	21.4	6	258	110	3.08
Hornet Sportabout	18.7	8	360	175	3.15

7.3 Tables

There are a number of packages that make nice tables in R Markdown. We will use the `knitr` package for this example.

The easiest way to make a table in R Markdown is to make a data.frame with all the data (and column names) you want and then show that data.frame (there are also packages that can make tables from regression output though that won't be covered in this book). For this example we will subset (which we'll cover in Chapter 10) the *mtcars* data (which is included in R) to just the first 5 rows and columns. The `kable` function from the `knitr` package will then make a nice looking table. With `kable` you can add the caption directly in the `kable()` function. The option `echo` in our code chunk is not set to FALSE here so you can see the code.

```
library(knitr)
mtcars_small <- mtcars[1:5, 1:5]
kable(mtcars_small, caption = "This is an example table caption")
```

For another package to make very nice looking tables, see this guide[4] to the `kableExtra` package.

7.4 Footnotes

In your writing, you'll often have sentences that you want to include but are auxiliary to your main point (or, frequently, to include links to specific

---

[4] https://cran.r-project.org/web/packages/kableExtra/vignettes/awesome_table_in_h tml.html

resources such as a website where you got data from). In these cases you'll want to include that info as a footnote, which is a section at the bottom of the page for this kind of information. To create a footnote in R Markdown, you use the carrot ^ followed immediately by square brackets []. Put the text inside of the [] and it'll print that at the bottom of the page. Code for a footnote will look like this: ^[This sentence will be printed as a footnote.]. In cases where you have a very long footnote it may extend to the next page and will be again at the bottom of the page. Look down at the bottom of this page to see the footnote (in a PDF or Word Doc, the footnote will be on the page you create it on, however since websites are just one long page without breaks, this footnote is at the very bottom of this entire page).[5] When you use a footnote, you'll usually put it immediately after the punctuation of the sentence it should be after. Note that footnotes are numbered so you can identify them. There's a blue superscript 1 where we make the first footnote. If we make another footnote, it'll be numbered sequentially, such that the next one is 2, the next is 3, etc.

If you're familiar with LaTeX you can use LaTeX code such as \footnote{} where the text goes inside the {}. But note that citations (which we'll learn in Section 7.5) won't work properly in the footnote if made this way. You can use LaTeX code - and use LaTeX packages - in R Markdown if you'd like and it'll operate (in most cases) like normal LaTeX.

7.5 Citation

In academic research you will need to cite the papers that you are referencing. R Markdown has a built-in way to cite papers, though it's a bit of a process to get everything setup. You'll need the citation data in BibTeX format and we'll walk through the steps from finding an article that you want to cite to citing it in your R Markdown file. First, a brief overview of what kinds of citations you can use. There are two types of citations you can use, in-text and parenthetical. You'll use in-text citations when you want to have the author names be in the text, and parenthetical citations when you want everything to be in parentheses.

Note, there may be other ways to get the citations in the right format; I'm just showing you one way to do so. For this example, we'll use the article "Using NIBRS data to analyze violent crime" by Brian Reaves that was published in 1993. We'll walk through the process from finding the article on Google Scholar to citing it in your paper. First, from Google Scholar we'll search for the article title.

---

[5]This is an example of a footnote.

Using NIBRS data to analyze violent crime

☐ **Using NIBRS data to analyze violent crime**
B Reaves - 1993 - books.google.com
With 1991 data, the Uniform Crime Reports (UCR) program of the Federal Bureau of
Investigation (FBI) began moving from summary counts to a more comprehensive and
detailed reporting system known as the National Incident-Based Reporting System (NIBRS).
By 1982 the Bureau of Justice Statistics (BJS) had already provided over $11 million to
States to establish centralized State-level UCR programs. In a 1985 report, an
𝄢 Cited by 31   Related articles   All 5 versions ≫

Showing the best result for this search. See all results

This returns all articles that meet your search criteria. Since we're searching
for a specific article title, we only get one result. The result shows some basic
info about the article - title, date, name, abstract. Below the abstract are some
important things. First, and circled in blue in the above photo, is a link that
looks like quotation marks. This is what we'll click on to get to the BibTeX
citation. While not necessary for citation, the next two links may come in
handy during your research. "Cited by 31" means that 31 published (in some
format that Google can locate, not necessarily peer-reviewed) articles have
cited this article. If you click the link it'll open up a Google Scholar page with
all of these articles. This is a good way to find relevant literature. Clicking
'Related articles' does the same thing but with articles that Google Scholar
deems similar, not necessarily articles linking to the one you're looking up.

But back to the quotes link circled in blue. Click this and it'll make a popup,
shown below, of ways to cite this article is various formats. We'll have R
Markdown automatically generate the citation in the format we want so we
don't need to worry about this. Instead, click the BibTeX link at the bottom
left.

as the National Incident-Based Reporting System (NIBRS).
atistics (BJS) had already provided over $11 million to
ate-lev
s All

See al

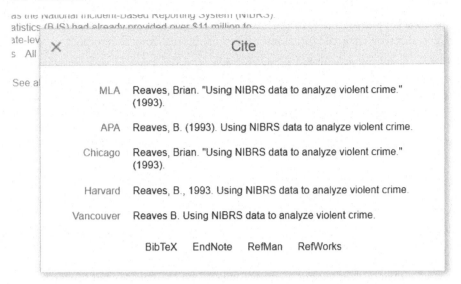

When you click it, it'll open up a new page with that article's citation in BibTeX form, as shown below. This basically is just a way to tell a computer how to cite it properly. Each part of the citation - author, year, title, etc. - is its own piece. Take a close look at the section immediately after the first squiggly bracket, "reaves1993using". This is how you'll identify the article in R Markdown so R knows which article to cite. It's essentially the citation's name. It's created automatically by combining the author name (first author if there are more than one author, publication year, and part of the title). You can change it to whatever you want it to be called.

```
@misc{reaves1993using,
   title={Using NIBRS data to analyze violent crime},
   author={Reaves, Brian},
   year={1993},
   publisher={US Department of Justice, Office of Justice Programs, Bureau of Justice~…}
}
```

Note at the end of the publisher section are the characters "~...". This looks like a mistake made by Google Scholar so we'll need to delete that so it isn't included in a paper we use this citation in. When using Google Scholar, you'll occasionally find issues like this which you'll need to fix manually - a bigger issue is apostrophes or other punctuation may copy over from Google Scholar weirdly (meaning that it copies as a character that your computer, and thus R Markdown, doesn't understand) and needs to be rewritten so R Markdown will run. You can rewrite it by just deleting the punctuation and typing it using your keyboard. This isn't always an issue so don't worry about it unless you get an error with the citations when outputting your document.

Below is the citation included in my .bib file, and the start of another citation also included in the file. A .bib file is basically a text file that programs can read to get citation info. You'll have all of your citations (in the BibTeX format) in this one file. To make a .bib file you can open up a text document, such as through the Notepad app in Windows, and paste the BibTeX that you've copied from Google Scholar. Save this file as a .bib extension (by renaming it filename.bib) and you'll have a usable .bib file.

Note that I have the word NIBRS surrounded by squiggly brackets {}. That is because by default R Markdown (and other citation generators such as Overleaf) will only capitalize the first letter of the title or the first letter following a colon. Since NIBRS is an abbreviation and should be capitalized, I put it in the {} to force it to remain capitalized. This is often a problem with abbreviations or country names (such as United States) in the paper title. Since all citations you use for a project should be in a single .bib file, you can see the start of another article citation below the Reaves citation.

```
@misc{reaves1993using,
    title={Using {NIBRS} data to analyze violent crime},
    author={Reaves, Brian},
    year={1993},
    publisher={US Department of Justice, Office of Justice
Programs, Bureau of Justice}
}

@article{jain2000recruitment,
    title={Recruitment, selection and promotion of visible-minority
and {A}boriginal police officers in selected {C}anadian police
services},
    author={Jain, Harish C and Singh, Parbudyal and Agocs, Carol},
    journal={Canadian Public Administration},
    volume={43}
```

To use citations from your .bib file, add `bibliography: references_file_name.bib` to the head of your R Markdown file. If your .bib file isn't in the R Markdown file's working directory, as my example below is not, you'll need to include the path in the file name.

```
title: "Ambient Lighting and Arrests: Evidence from a Natural Experiment"
author: Jacob Kaplan^[Department of Criminology, 483 McNeil Building, University of
Pennsylvania, jacobkap@sas.upenn.edu. For helpful comments and suggestions, thank you to
Sara-Laure Faraji and Kristina Block. All remaining errors are my own.]
bibliography: C:/Users/user/Dropbox/Penn/PhD/Research/global_references.bib
```

Now that we have the citation in BibTeX format, have put it in our .bib file, and have told R Markdown where to look for that file, we are ready to finally cite that article. To use a citation we simply put the @ sign in front of the citation name (in our case "reaves1993using") so we would write @reaves1993using. This will give us an in-text citation, with the author name in the text and the year in parentheses. Adding a - right in front of the @ will cause the citation to show just the year, not the author's name. You'll usually want to use this if you've already named the author earlier in the sentence. Generally we will want parenthetical citations, with both the authors and the year in parentheses. To do this, we put the citation inside of square brackets like this [@reaves1993using]. If we're citing multiple articles, we separate each citation using a semicolon [@reaves1993using; @jain2000recruitment].

Here's what the results look like when citing that Reaves article, see the image below for what this looks like just as code.

(Reaves, 1993)

Reaves (1993)

-Reaves (1993)

(1993)

(Reaves, 1993; Jain et al., 2000)

If you use a citation that isn't in your .bib file, R Markdown will show a question mark, indicating that you made some mistake.

(?)

```
[@reaves1993using]

@reaves1993using

-@reaves1993using

[-@reaves1993using]

[@reaves1993using; @jain2000recruitment]

If you use a citation that isn't in your .bib file, RMarkdown will present three question
marks in place of the citation.

[@wrongCitation]
```

When you use citations, R will automatically put the reference section at the very end of the document. Two LaTeX commands may be useful here. \clearpage makes a new page so your reference section isn't on the same page as the conclusion. \singlespace makes the reference section single spaced if your document is set to be double spaced. Put these commands at the very end of your document so they only apply to the reference page. You don't need to do anything other than write them (for easier reading, make them on separate lines) at the end of the R Markdown file. If you want to make the references go in another part of the paper (e.g. after tables and figures), just put this code at the place in the paper where you want to reference section to go: `<div id="refs"></div>`.

7.6 Spell check

R Markdown does have a built-in spell checker (the ABC above a check mark symbol to the left of the Knit button) but it isn't that great. I recommend that you export to Word (or open up the PDF in Word if you prefer using PDFs) and using Word's superior spell checker.

7.7 Making the output file

To create the Word or PDF output click `Knit` and it will create the output in the format set in the very top. To change this format click the white down-arrow directly to the right of `Knit` and it will drop-down a menu with output options. Click the option you want and it will output it in that format and change that to the new default. Sometimes it takes a while for it to output, so be patient.

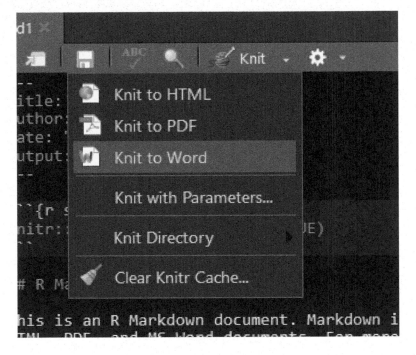

8

Testing your code

This chapter covers how to write code that tests other code. It's especially useful when you write complex functions but is also useful for work such as PDF scraping or webscraping where you know the right answer (by looking at the PDF or web page yourself) and want to be sure your scraping code did the scrape correctly. However, in most cases when programming for research you won't formally test your code - though you should be checking if everything makes sense and rereading your code to look out for errors (such as typos or using the wrong data). If you've never programmed before, I recommend that you skip this chapter entirely (or read it but don't feel pressure to understand everything) and return to it after you've finished the rest of the book.

8.1   Why test your code?

As you write code, you will inevitably make mistakes. There are two main types of mistakes with coding - those that prevent code from working (i.e. give you an error message and don't run the code) and those that run the code but give you the wrong result. Of these, the first is probably more frustrating as R tends to give fairly unhelpful error messages and you'll feel you hit a roadblock since R just isn't working right. However, the second issue - code is wrong but doesn't tell you it's wrong! - is far more dangerous. This is especially true for research projects.

Let's use examining whether a policy affected murder as an example. In the example data set below, we have two years of data for both murder and theft, and we'll say that the policy changed at the start of the second year. If we want to see if murder changed from 2000 to 2001, we could (overly simply) see if the number of murders in 2001 was different from the number in 2000. And since the data also has theft, we'd want to subset to murder first.

```
example_data <- data.frame(
  year = c(2000, 2000, 2001, 2001),
```

```
  crime_type = c("murder", "theft", "murder", "theft"),
  crime_count = c(100, 100, 200, 50)
)
example_data
#    year crime_type crime_count
# 1 2000     murder         100
# 2 2000      theft         100
# 3 2001     murder         200
# 4 2001      theft          50
```

To see if murder changed, we can subset to the rows where the crime is murder, and then print out the year and crime_count columns to see if there is a change. So our code will be `example_data[example_data$crime_type == "murder", c("year", "crime_count")]`. Below I've accidentally only put one = instead of two, this will give us an error and not give any other results. Helpfully, the error message tells us that there's an error with the = sign, though not what that exact error is.

```
example_data[example_data$crime_type = "murder", c("year", "crime_count")]
# Error: <text>:1:38: unexpected '='
# 1: example_data[example_data$crime_type =
#                                          ^
```

Now I've made a different mistake. Here, instead of ==, I've written != which is the opposite of what we want - it'll return all rows that do **not** equal "murder". Now it looks like the policy cut murder in half when in actuality the policy doubled murders! Since we don't print out the type of crime in the output, we wouldn't catch this from the output alone.

```
example_data[example_data$crime_type != "murder", c("year", "crime_count")]
#    year crime_count
# 2 2000         100
# 4 2001          50
```

You may think this is a silly example that is unrealistic. And it is to a degree, it's just one line of code that we're using to evaluate an entire policy. Now think about how you would actually evaluate a policy using data that you're familiar with. Now the code is going to be much more complex. Your code may be hundreds of lines long, deal with multiple data sets that must be joined together, and involve a number of relative subjective (though must be defensible) decisions as to how to deal with your data (e.g. what crimes constitute violent crime, what time unit to analyze), and some of the code may be written by other people who you are collaborating with. The increased complexity with a real analysis increases the likelihood that errors will occur - and even small issues such as an incorrect subset can have large impacts on your results.

So, how do we properly test our code? There are two main methods that I'll refer to as informally testing and formally testing. The formal method will be using something called "unit tests" that we'll discuss in the next part of this chapter.

Informal methods are what you've likely been doing already. Essentially, just looking at your data and trying to see if it "looks right". This includes stuff like printing summary statistics (using `summary()`) of important variables and making simple graphs to look at the data. If something is wrong, exploring the data is a fairly good way to discover it. For example, if you are looking at arson data from the FBI, you may find (as this is actually in the data) some cities with millions of car arsons in a month. This is clearly wrong so you know there's an issue - in this case, an issue with not subsetting out obvious outliers. Knowledge about the topic and the data are also important in this approach. If you are familiar with a given topic and your results are similar to that of past studies, that's a good sign that you did things right.[1]

You can also take this kind of approach when testing functions - which ideally are the way you write code. For example, if you have a function that takes a number and returns that number + 2, you can test it by checking a few cases. If you input 2, you expect 4. If you input -2, you expect 0. Do this a few times and you can be more confident that the function works properly. Now imagine a function that's more complex - one that calls a different function and uses the result of that function. If you change the underlying function, you'll need to check both that function and the function that calls it.

As you have more intertwined pieces in your code, this gets more and more complex. It also takes a lot more time as you'll have a lot of code that only checks a function and will have to run it line by line to see if there's an issue. At this point, relying on informal methods becomes unfeasible and you'll want to use unit tests, a formal way to test your code. Note, however, that this is far better suited for checking functions than for checking data, though it is possible to some degree. We'll discuss formally testing data in Section 8.2.3.1.

---

[1]However, make sure that you don't look less closely just because the results are the way you expect. Past results may be wrong, or you can have a new finding, so make sure to avoid complacency just because you like the results

8.2   Unit tests

A unit test is simply a conditional statement where you have some input, usually a function with some parameters set, and state what you expect the result to be. You are saying "I expect that if I do X, I will get Y". And if you get a result other than what you expected, R will tell you. In R, you can make a number of unit tests and have R run them all at once and inform you of which ones failed. Each unit test is just a function in R that is specifically for checking whether other functions - or other code or data - are correct. They operate just like a normal function. To use unit tests, we'll use the R package testthat which has a number of functions that make unit testing easier, and we'll use some keyboard shortcuts in RStudio that also improve the ease of testing.

Please note that these shortcuts will only work if you're working in an R Package, a normal R Project won't work. An R Package is a special type of R Project, which you can make by following the steps in Section 5.2 and choosing "R Package" instead of "New Project" in the Project Type panel. An R Package is essentially an R Project with the goal of creating a package in R, though there's no requirement that we actually make a package. We can treat it as a normal R Project but use the added testing tools.

If you don't have testthat installed, do so using install.packages("testthat"). For more information on the package, please see the package's website[2].

```
install.packages("testthat")
```

```
library(testthat)
#
# Attaching package: 'testthat'
# The following object is masked from 'package:tidyr':
#
#     matches
# The following object is masked from 'package:dplyr':
#
#     matches
# The following objects are masked from 'package:readr':
#
#     edition_get, local_edition
```

---

[2]https://testthat.r-lib.org/index.html

In testthat, every function follows the same expect_ format where a type of conditional statement follows the __. For example, expect_equal() checks if two values are equal, expect_named() checks if the name of a data set is correct, and expect_silent() makes sure that the code that's run doesn't return any warnings, messages, or errors. To use this technique for our above example of the function that adds 2 to an inputted number - which we'll call add_2() - we can use some expect_equal() functions. If we input 2, we expect 4. So we'd write expect_that(add_2(2), 4).

```
add_2 <- function(number) {
  return(number + 2)
}
expect_equal(add_2(2), 4)
```

Above is the code that makes the add_2() function and one unit test checking it. It doesn't output anything. That is good. When a test passes, there is no information; when it fails, the function will output a message that it failed. Below is another test, this one intentionally wrong to show what happens when a test fails.

```
expect_equal(add_2(2), 5)
# Error: add_2(2) not equal to 5.
# 1/1 mismatches
# [1] 4 - 5 == -1
```

It gives an error, telling us that the result of the add_2(2) function does not equal 5. Helpfully, it also shows us how much of a difference there is between what we expected and what we got. Note that it says "1/1 mismatches". That says that all of the expected values - we only expect one value here - are incorrect. If we expect more than one result, such as if we expect a function to return a vector, it will check each value and say exactly which ones (in the order we have the resulting vector) are incorrect. This is helpful when diagnosing exactly which part failed.

There are a few different ways to run the unit tests. First, you can run them like a normal R script by running each line directly. This is fairly inefficient and loses some of the benefits built into RStudio for testing. Though in this case you do **not** need to be using an R Package, you can just use a normal R Project or just use normal R and the code will work fine.

The next way is to use the test_dir() function from the testthat package where you enter the folder directory in the parentheses and it runs every test file in that folder. It's easier to simply use the test_path() function which inserts the correct folder, assuming that you didn't move folders around after use_test()

(which we'll discuss below) created them. So you'd write test_dir(test_path())
and it would run all of your tests. This also prints out a nice summary of the
results for all of your tests combined, showing the number of tests that passed,
that failed (i.e. didn't pass), that returned warnings, and that were skipped
(you can force R to skip some tests which is useful when you change one part
of the code and know those tests will fail but still want to test other parts).

You will also need to run all the functions or load all of the data that the tests
check, otherwise you'll get an error since R doesn't implicitly know that the
functions/data exist. You can run this the normal way by highlighting and
running the function (or the code to load data) in your R script or use the
shortcut Control+Shift+L (Command+Shift+L on a Mac, the L stands for
"Load") which will run every R file in your project (and every line of each
file).

The final way is to use the keyboard shortcut Control+Shift+T (Com-
mand+Shift+T on a Mac, the T stands for "Test") which will load all of
the files in your folder and then run all of the tests. It's a quicker way of
doing the above method. However, this shortcut only works when using an R
Package, not a normal R Project.

```
> test_dir(test_path())
√ |  OK F  S | Context
x |   3 1    | test

test-test.R:5: failure: multiplication works
20 * 2 not equal to 41.
1/1 mismatches
[1] 40 - 41 == -1


== Results ==
OK:       3
Failed:   1
Warnings: 0
Skipped:  0

I believe in you!
> |
```

8.2.1 Modular test scripts

Before getting into exactly how to write a unit test correctly, we'll talk about
organizing each testing file. As with your normal R script, you can have separate testing scripts (a testing script is a normal R script which people use
specifically for testing code but doesn't actually function any different) for
each major part of the code that you're testing.[3] As with the R scripts for

---

[3]For more info on having separate R scripts for each major section of your code, please
refer to Section 5.3

your code, this is simply a way to organize your work, and doesn't affect the testing. Below is an image showing the files I use to test the US Border Patrol scrapers. I have one file per PDF that I scraped.

Note where the folder depicted above is located. It's in a folder called "test-that" in the "tests" folder in the main project folder that I called "border-patrol." We'll use a helpful function from the usethis package to organize our test files and generate them automatically. If you haven't installed this package already, do so using `install.packages("usethis")` and then load it with `library(usethis)`.[4]

You can use the function `use_test()` from the usethis package to create a test file inside your R Project. This will automatically create the necessary file and folders (if not created already) so you don't have to do any more work. Run this function by putting the name of the test file you want to create (in quotes) in the parentheses. It will open the test file in the Source panel (shown in the top left). In the example shown below, I wrote `use_test("test")` to make a new file called "test". In the Source panel, the file is called "test-test.R," which is just because usethis will automatically add "test-" to the name of any test file name you make. `use_test()` will also generate an example of a test, which you can modify (or delete entirely) to suit your own needs.

---

[4]The usethis package is an extremely helpful package that automates a lot of work that you would do primarily for R package development so if you go down that route I recommend exploring the package more through its website https://usethis.r-lib.org/index.html).

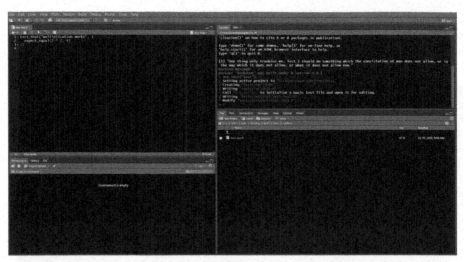

The first file in the testthat folder is called setup.R, which is a file that will automatically run first when you run a test script through R or using RStudio's keyboard shortcut. This file is where you run some code that is used during the tests. In my setup.R file I made several vectors, which I use during the tests to subset the data. You won't always need to have a setup.R file, but it's useful when you want to run the same code beforehand for multiple different test scripts.

8.2.2 How to write unit tests

We'll start by looking at the default test example made when using `use_test()` to understand the organization of a test file before getting into an example of actual tests. In the image below, there are really two pieces. First, we have the actual test on line 2 - `expect_equal(2 * 2, 4)`. This is saying, I expect 2 * 2 to equal 4, and R will check if that is true. All of your tests will be in this format, just for a specific result from a specific input. Now let's look at the code surrounding that line - `test_that("multiplication works", {})` where the `expect_equal()` line goes inside the {}.

The `test_that` code is basically a form of organization within a test file to group similar tests together. In this case it is grouping all of the tests that check if "multiplication works," though we only have one test written. Below I've added three new tests to this "multiplication works" testing group. To run this code, I can either run each `expect_equal()` individually (remember to run `library(testthat)` beforehand or it won't run) or run the entire `test_that()` group at once. You can do this by either highlighting it all and running it or selecting either the top or bottom line (which has the squiggly brackets) and running that line - the entire thing will run.

The benefit of this is that when you run all the tests you write (and you'll often have many test groups and more individual tests than shown here), if a test in a group fails, it will tell you exactly which group failed (based on the name of the group which you specify - here, "multiplication works"). Note that the final test in this example is incorrect, and in the Console panel on the right it says that "Test failed: 'multiplication works' to tell you where the test failed. The test groups aren't necessary, but they make it easier to organize your tests.

As an example of actual tests, we'll go over the tests that I wrote when I first scraped the US Border Patrol data that we will scrape in Chapter 22. This test file is organized almost identically to the example one shown above. At the start I have some code that loads the data that I will test - this isn't in the setup file since the code is for this specific test script (though it could be in the setup.R file and the results would be the same). While most tests check the result of functions, here I am checking the data that is outputted by the function, and not rerunning the function for each test. I do this because the function that scrapes the PDF is relatively slow to run and I have many tests, but putting the function that gets the data in the test directly will give the exact same results. Then there are several test_that() groups with some expect_equal() tests inside each.

Since these tests are checking if the code is scraping the PDFs correctly, I determine the expected result by looking at the PDFs and writing down what the values should be (be careful, this must be done by hand but that can mean you mistype - so double-check your work!). We'll use the test on lines 21-22 as an example. Here I am asking if the values in the "cocaine_pounds" column, for rows where the sector is "coastal border" are equal to the values c(6843, 1701, 3169, 1288, 6884, 20, 709, 5962, 989). If they are, then the scraping was correct (at least for this part of the PDF) and the code worked. In this case I checked every value that meets the two conditions, but that's just because there were relatively few values. If I had many values that meet those conditions (i.e. many rows of data in that column), I would just check a small number of them.

8.2.3 What to test

Now that we've gone over how to make unit tests, let's talk about what to test. When testing functions, you generally want to test every possible parameter in the function, and a variety of inputs. In particular, try to think of ways that the function could be used incorrectly and write a test to catch that. For example, our add_2() function will fail if a string (e.g. "2") was inputted instead of a number, so you'll want to add a test for that. You'll also want to make sure that inputting something other than simply a single number, such as a vector of numbers, works as expected. Basically, you want to be thorough and cover all of your bases.

Writing unit tests is one of the most time-efficient things you can do since it helps you avoid making costly (in time and in getting wrong results) mistakes. But don't spend too much time writing tests. If you've tested that add_2(2) equals 4, no need to test that add_2(3) equals five since you're essentially testing the exact same thing. And consider your audience (even if that is only you). If you know that add_2() will only be used by people who know better than to input a string, there's no need to test for that. In general, I think it's always better to have more tests than fewer, but consider whether writing that test is a good use of your time. This is something that you'll learn with experience so it's better to have too many tests when you're first using R than too few.

8.2.3.1   Tests for research projects

When you use R for a research project, you'll usually take data that someone else collected, or scrape it yourself, do some work to clean this data (e.g. subset or aggregate the data, standardize values) and then run a regression on it. In these cases there are relatively few opportunities to use unit tests to check your code. Indeed, the best checks are often content knowledge about the data and examining the results of your analysis to see if it makes sense and fits prior literature.

While testing is most commonly used for functions, you can use it to test data. Writing tests for research data is best if your code is scraping the data (webscrape or PDF scrape) and you want to verify that it is correct, or if you expect the data to change and want to ensure that it is still correct (while exact values will change, you can check broad categories such as whether certain groups are included). For example, if you know that you only want to look at a certain state, you can write a test that expects the only state in the data to be the one you're analyzing. This way, if you add more data, such as a new release of that data set, the test will catch if there's any other state that you may have forgotten to remove after adding the new data. If you're sure that you will only use a particular data set that never changes, you're better off just writing code in your main R script (or a specific script for checking the data) to do these checks rather than dedicated tests.

8.2.3.2   Tests for data collection

Our example in this chapter was tests for a data collection process - in our case, PDF scraping - so we've already seen how to test code for gathering data. We'll still talk briefly here about what kind of tests - and how many tests - you will want for this type of code. In normal tests, you don't want to test the exact same thing multiple times (for example, if you test that $2 + 2 = 4$, you don't need to test that $2 + 3 = 5$). This is different when it comes to testing code that collects data from a source, such as through PDF scraping or webscraping.

When testing data collection code, you want to be far more thorough, retesting something in multiple ways. This is because small differences in the data you are scraping may affect the code at different parts of the scrape. For example, imagine a PDF with ten pages and a single table on each page. On the first page there five columns but on the next nine pages there are six columns. If you test only data from the first page you'll miss all of the pages where there are six columns instead of five - and where your code to scrape it is probably wrong.

I've often experienced PDF or webscraping where some parts of the data are just "weird" and cause the code to scrape it incorrectly - but often not tell me that there's an issue. So to catch this you'll need far more tests than normal. I prefer to choose a few random pages (more if the PDF/website is longer) and test random rows and columns since that'll give a good coverage of the results. In addition, I look at the PDF or website and try to see if there's anything atypical about a certain part; if there is, I test that specifically. It's easy to over-test (and that's better than under-testing) this kind of work, but there are rapidly diminishing returns. So test comprehensively but not at the cost of having too little time to work on code - again, this is something that requires experience and doesn't have a hard rule on what constitutes too much (or too little) testing.

8.3   Test-driven development (TDD)

We'll finish this chapter by talking about test-driven development (TDD), a philosophy in programming where you write the tests first and then write the code that meets these tests after. This is really an extension to the discussion in Chapter 5 of planning out your project before you start. In Chapter 5 we talked about writing out every step of the project and hand sketching all the figures or tables that you intended to have. With TDD, you write tests for all of the functions you intend to write (and any variations of parameters or inputs for these functions) or data you intend to gather/clean.

Test-driven development is a useful tool to make you really think about the functions that you need to write, and how they interact with each other. This is an excellent way to identify potential issues (I've often realized while writing tests that the approach I was going to do wouldn't work) before you start on the code. However, for this same reason, it is a fairly advanced topic since you need to know exactly (or, mostly) what you need to do, and the likely problems that each approach will face. For that reason, I recommend holding off on using TDD until you're fairly experienced with R or programming in general.

9

Git

This chapter covers Git, which is a way to have version control for your code - like a programming version of Dropbox, but with a few added features. This is relatively advanced material and isn't necessary for using R. However, when you're dealing with complex projects or with multiple collaborators it is helpful to use. Given that this material is relatively advanced, feel free to skim or skip this chapter entirely, and come back to it when you think you need it - which will likely be after you finish the rest of the book.

9.1   What is Git, and why do I need it?

As you write R code you will - I hope! - save your R script from time to time to avoid losing any code you've written if you close R or shut down your computer. This is important as it'll save everything you've done locally, but if your computer crashes you'll want your work to be backed up elsewhere. While you should have something like Dropbox or Google Drive that keeps backups of your work, here we'll talk about Git, which is a version control software that gives you much more control (but requires more work) of the saved work than from something like Dropbox.[1]

Before getting into exactly how to use Git, we'll talk first about what it is and how it'll help your work. Git is also a very powerful and complex tool so this guide is going to be touching just a small - but useful to most researchers and R programmers - part of it.

With backup software such as Dropbox, it'll save your work very frequently - so frequently in fact that I sometimes turn off Dropbox when I write R since it keeps interrupting me by saving at the moment I'm typing, which stops the typing. The following image is the Dropbox page for some R code that I've been working on to scrape Covid data. Notice the timestamps - 4/5 of

---

[1]This came in handy for me as somehow one of my dissertation papers written in R Markdown became empty a couple of months before my defense, and I couldn't undo that change. My Dropbox backup was older than my Git backup so having Git was a real time-saver.

them are within one minute, showing how often Dropbox is saving changes. This is useful if I need the most recent update - or to share the most recent version with a collaborator. Here's the big issue - and the one that Git solves - I have four versions within a minute of each other: what's the difference between them? Dropbox is saving automatically and doesn't indicate how they're different (clicking on the file shows the complete file, not differences relative to some previous version), which means if I mess up some code a while ago, I can't easily see which version is the one that works. With Git you can wait until you've made enough changes to decide that these changes merit a new "version" of your work.

If you've ever used the track changes feature on a Word Document, the concept is similar. When you have this setting in a Word Document every time you (or anyone else) makes changes in that document, those changes, who made them, and when they occurred, is tracked. This makes it easy to see exactly what part of the file was changed and to undo that change if necessary. Below is an example of this feature on one of my drafts on Overleaf (basically a way to collaborate using LaTeX, which is similar to R Markdown). You can see each change that my co-author made in the draft in the purple changes in the main part of the photo. The parts that were rewritten or added are highlighted in purple while the parts that were deleted are crossed out. What is shown in purple isn't all of the history of changes for this paper. If you look at the part on the right, highlighted in green, it shows what files were edited, by whom, and at what time. If you don't like a change - or in R's case more commonly, broke some code by accident - you can go back in the history of changes and return to an older version.

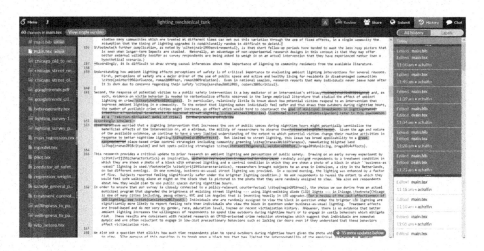

The way that R - and many other programming languages (and technically you can use this for any file or folder) does this "version control" is through Git.

9.2 Git basics

There are four main processes you need to know for a basic understanding of Git: checkout, add and commit, push, and pull. This chapter will explain how to use Git through buttons on RStudio so you don't necessarily need to know these commands in Git, but it's useful to know enough to talk about them and ask questions if needed.

We'll use the example of getting a book from the library to walk through using Git. The steps for this are simple, we go to the library, pick a book we want, check it out from the librarian, read it, and eventually return it. Using Git adds one wrinkle to this: we will want to write in the book and see what other people write too. Of course, when the book is checked out, no one else could write in our version, and no one can see what we write. So anything we write has to be done before we return the book to the library, then we check out the book again to see what other people have written. When we want another book, we simply redo these steps.

Library Steps	Git steps	Git code
Go to library		
Find book and check out book	Clone (usually will just be done once per project).	Git clone *path to repo, can be GitHub link*
Read or write in book	This is done in R, not in Git	No Git code, this is going to be whatever code we write in R. Also includes any outputs such as making a graph that is saved, R Markdown outputs like a PDF, or even new R files.
Return book	Add and commit Push	Git add . Git commit –m "message indicating what we wrote" Git push
Check out book again (to see what other people have written in it)	Pull	Git pull

Another way to think about commit vs push is that of writing an email. When you write an email, you're essentially editing a blank document by adding the words of the email. When you save (but don't send) the email, you are making a commit (essentially "committing" or promising to make a change). When you send the email you are making a push (taking something that you have written and changed and sending it to the main repository). While emails let you correspond directly between two or more people, how Git works is like sending the email to a central server (or a post office) and anyone who wants to read it has to go there. And when someone reads it and responds, their email also goes to this central server. You have to go there to get their response (called a "pull" in Git terms), which is essentially an addition to your initial email.

9.3   Using Git

While you can use Git like writing R code (though the syntax is not that similar to R), RStudio has built-in buttons that work instead of writing code yourself. We'll go through these buttons and not discuss any Git code beyond the small amount needed to link your project to GitHub, a website that is like Dropbox for code.

9.3.1 Setting up Git

To install Git on your computer install Git for Windows[2] for Windows computers and Xcode[3] for Mac computers. If you're on a Linux operating system, see here[4] for how to install Git. For more help I recommend this chapter[5] of Happy Git and GitHub for the useR, which covers installing Git.

You'll now need to tell Git some identifying information about yourself so that whenever you make a commit, Git will know who you are. We will use a function from the usethis package to do this. The only information we need is your name (or nickname, just something so collaborators know that it was you who did a certain commit) and email address (below you'll set up an account on GitHub - use the same email address there as here). We'll use the function use_git_config, which has two parameters - user.name and user.email, which take strings with your name and email, respectively.

```
library(usethis)
use_git_config(
  user.name = "Your name",
  user.email = "email_address@gmail.com"
)
```

Once you have Git installed, you'll need to enable it through RStudio. To do this, go to Tools and click Global Options. Then go to the Git/SVN tab and check the "Enable version control interface for RStudio projects" checkbox. The final step here is to click the first Browse button and navigate to where you installed Git on your computer. Select the Git file (on a Windows computer this will be within the larger Git folder) and then hit OK to close the popup.

---

[2] https://gitforwindows.org/
[3] https://git-scm.com/download/mac
[4] https://git-scm.com/download/linux
[5] https://happygitwithr.com/install-Git.html

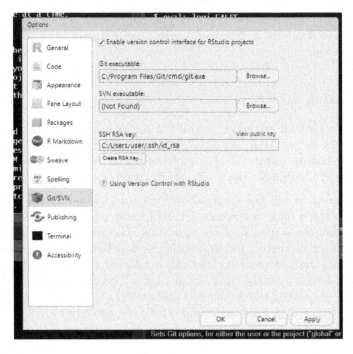

9.3.2   Setting up GitHub

We'll be using GitHub to host our Git commits. To use GitHub, please make an account on their website.[6] There are several types of accounts[7] at various monthly costs, but you only need the free version. This gives you an unlimited number of public and private repositories (sometimes shorthanded to 'repos') - these are basically R Projects (you can use any language when it comes to using Git and GitHub, not just R).

A public repository is one that anyone can look at on GitHub, download the code/files, and make any changes they want (though if they want to make changes to your repository they need to make a change request that requires your approval; it is not automatic). This is good for projects where you want others to collaborate on or to showcase your work. A private repository is the same thing, but only people you approve can view, download, and work on your repository. This is good for when you don't want the code to be public (e.g. code for an employer or dealing with sensitive data, such as people's personal information). I tend to keep my research work private until the paper

---

[6]https://github.com/
[7]https://github.com/pricing

is published and my data work public since I want people to notice it and find bugs.[8]

Once you've made an account on GitHub, you'll need to create a repository there to connect to your R Project. You can do this through the GitHub home page as shown in the following image. This page is my own homepage and shows several of my current repositories on the left (note the ones with a golden lock to the left, these are the private repositories which are only accessible to people I permit), a list of updates on other people's repositories that I chose to get updates from, and some suggested repositories that GitHub thinks I'd be interested in on the right. To create a new repository, click the green New button on the left side above the list of current repositories.

After you click the green New button, you'll go to a page when you set a name for your repository (this can be different from the name of your R Project, though I prefer to use the same name so I know exactly what project the repository is for), provide a short description, and choose if the repository should be public or private. You can also optionally add a README file, which is a longer form of description for what the code is and its purpose (basically a short manual for the project - often explaining how, not why, it works), and add a .gitignore file or set a license (which tells people who look at the project what they're allowed to do with it. For more on code licenses please see this excellent site.[9]) The .gitignore file is essentially a list of files or folders than you do **not** want to upload to GitHub. These last three choices are all optional. and if you don't do it now, you can do it anytime through R. Once you've made your choices, click the green Create Repository button

---

[8]You may disagree with my decision to keep research code private until publication - and for good reason. Doing this has the benefit of preventing people from scooping my (and my collaborators') work, but also makes it more likely to lead to bugs as there are fewer people looking at the code.

[9]https://choosealicense.com/

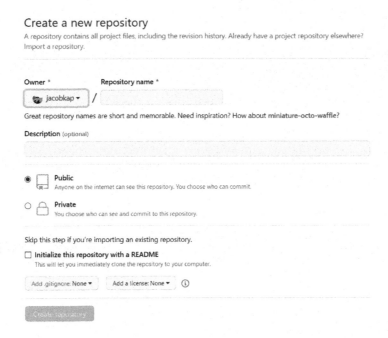

This will open up a new page with a bunch of code that you'll enter in R that connects your Git commits to this repository on GitHub. We'll get to this in a bit - for now, let's focus on those three buttons in the top right. These are for accessing or following other people's public repositories (you can technically click on them in your own repository, but there isn't much benefit to that apart from the first button).

The first button sets your notification settings for the repository. To change the notification setting, click "Unwatch" and then select what you want to be notified for. By default it is set to notify you of all conversations that occur. The main conversation will be when someone posts a message in the Issues tab where they tell you about an issue (or sometimes make a request for a new feature or just ask a question) about the code in this repo. With your own repositories, you'll want to be notified of all conversations so you don't miss anything. You can use this option on other people's repositories, and it will alert you of changes or conversations in that repo. This is useful when you want to know about updates (i.e. new features) on repositories that you're interested in (for example, I follow the testthat repo[10] so I know of any new versions of that package that may have useful features).

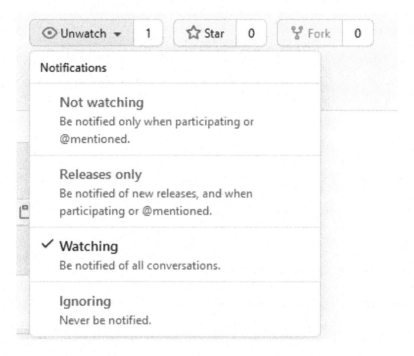

Stars are simply a way to favorite a repository, and you can see a list of all repositories that you have starred by clicking the profile button on the top right and going to "Your stars."

---

[10]https://gitHub.com/r-lib/testthat

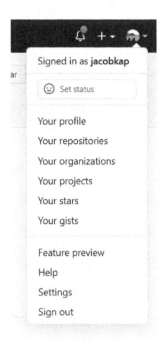

The final option is "Fork" which creates a new repository on your account that is a copy of the repository that you forked. You will occasionally want to fork other people's repositories - there isn't much benefit of forking your own as that's essentially just making a duplicate of your own work - and modify them to suit your needs. This is useful for two reasons. First, if you want to collaborate with someone - even if just to submit a fix to a bug you found (or a typo in this book!) - you can fork their repository, make the changes on your own R Project, commit the changes, and request that the original account accept your changes into the repository that you forked (called a "pull request").

This sounds very complicated to make what could be a simple change (and it is) so why bother? As you get more familiar with R and how R handles Git, this process won't take *too* much extra time so it's not that much of an additional burden. But the main advantage is that Git establishes much more structure than would exist otherwise, and helps protect the original creator's time. Consider that you found a bug in some of my code and sent me an email detailing that issue. This is probably the best-case scenario for you - it is quick to send emails. For me, that adds time to try to figure out what and where the bug is (describing it better would just take more time for you to write and me to read) and then to fix the bug. Even if you included the fix in the email, it would take me time to test it.

When using Git and GitHub, this process is far easier for the person receiving the changes (and while it is extra work because you must follow Git procedures, it can be somewhat easier as you won't need to explain as much). If you submit a bug fix to me through GitHub, I will immediately know what it changes as Git highlights all differences between my version and your fixed version, and I can set it to automatically run tests (see Chapter 8 for more on this) to make sure everything works. There are no longer any questions of what was changed, where the code was changed, or whether it passes all the unit tests (GitHub will run all unit tests and tell you if they pass). Everything is largely automated so accepting changes is a breeze. As you program and collaborate more, you'll increasingly be on the side of receiving changes to your code, so the balance between extra work as a submitter and easier time as a receiver of changes gets better.

In Section 5.2 we walked through making an R Project and selected the "Create a Git repository" box without explaining what that does. Clicking this box sets the R Project up to use Git so you don't need to do any other steps from the R side (but you'll need some steps to connect with GitHub). In the below section we discuss a simple way to connect your R Project to Git if you didn't check this box. If you plan on always checking the box - and have no unchecked R Projects that you want to use with Git, feel free to skip the following section.

9.4 Setting up Git on an already-made R Project

If you didn't tell RStudio to set up Git in your R Project, it's quite simple to do so through RStudio.

First, go to Tool -> Project Options. Then click the Git/SVN button that is second to the bottom to open up the Git options. This will open up a page that says "Version control system," which will be set to "(None)." Click this and set it to "Git."

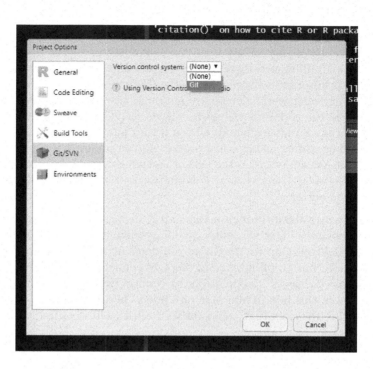

It will then ask if you want to set up Git for the current R Project. Say Yes.

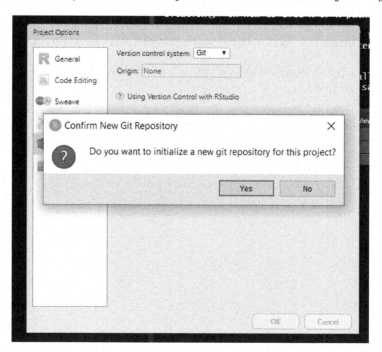

You need to restart RStudio for Git to work now, so click Yes.

Now if you look at the Environment panel you can see a new tab called "Git". We'll do all of the Git work in RStudio through this tab. You are now ready to use Git for this project.

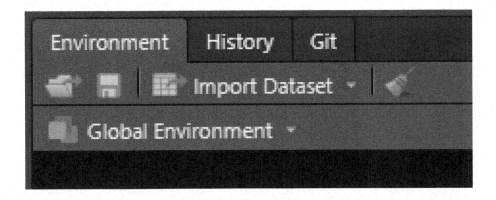

9.5   Using Git through RStudio

Now we have an R Project with Git ready, and a repo on GitHub to store the project files. We need a way to connect the R Project to the specific GitHub repo - for this, we'll return to that screen on GitHub with all of the weird code that starts with the word "git." We need to enter that code into R to connect the two. To do this, we need to use the Git Shell, which is basically like the Console panel but for Git. You can get to this by going to the Git tab, click on the More button, then click "Shell...".

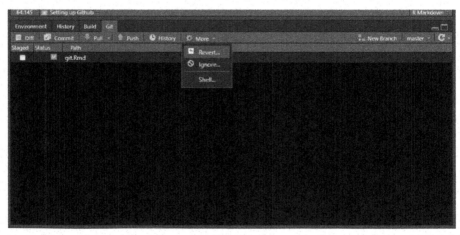

This opens up a popup almost identical to the Console panel. Here we can write the code (or copy it from GitHub) and hit enter/return to run the line. This is the only time we will be using actual Git code in this chapter (there is some benefit to learning the Git code rather than relying on the buttons in RStudio as it is much faster when dealing with large files or simply a large number of files to use the code rather than through RStudio - though I'm not sure why this is).

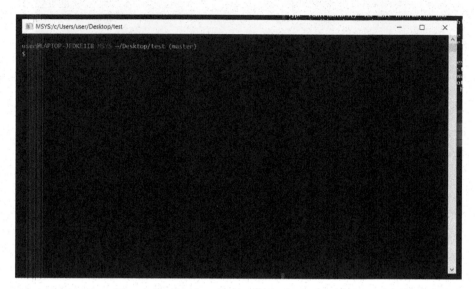

We will use the first chunk of code that's shown on GitHub - the one that starts with the bold text **"...or create a new repository on the command line"**. You can copy and paste all of the code (starting with the "echo" line and ending with the "Git push -u origin master" line) to the shell and hit enter or you can do it one line at a time.

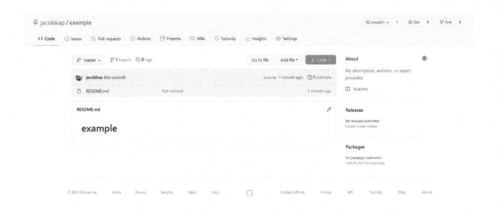

```
MSYS:/c/Users/user/Desktop/test                                    —   □   ×
$ git init
Reinitialized existing Git repository in C:/Users/user/Desktop/test/.git/

user@LAPTOP-JFDKE118 MSYS ~/Desktop/test (master)
$ git add README.md
warning: LF will be replaced by CRLF in README.md.
The file will have its original line endings in your working directory

user@LAPTOP-JFDKE118 MSYS ~/Desktop/test (master)
$ git commit -m "first commit"
[master (root-commit) 56da75d] first commit
 1 file changed, 1 insertion(+)
 create mode 100644 README.md

user@LAPTOP-JFDKE118 MSYS ~/Desktop/test (master)
$ git remote add origin https://github.com/jacobkap/example.git
fatal: remote origin already exists.

user@LAPTOP-JFDKE118 MSYS ~/Desktop/test (master)
$ git push -u origin master
Enumerating objects: 3, done.
Counting objects: 100% (3/3), done.
Writing objects: 100% (3/3), 220 bytes | 220.00 KiB/s, done.
Total 3 (delta 0), reused 0 (delta 0)
To https://github.com/jacobkap/example.git
 * [new branch]      master -> master
Branch 'master' set up to track remote branch 'master' from 'origin'.

user@LAPTOP-JFDKE118 MSYS ~/Desktop/test (master)
$
```

Refresh your GitHub page and you'll see that instead of code on the screen, it shows the files that you uploaded. In this case, I didn't make any files so it is largely blank, just a relatively empty README file. If this was a real project, you'd see all of the same files (except those you chose not to commit) as in your R Project folder. Your R Project is now connected to the GitHub repo so you can do the rest of the Git work on this project entirely through RStudio and will not need to touch the Git Shell again.

The below image shows my Git tab while working on this chapter and from an update to the Subsetting chapter. It has a list of all of the files that I changed since my last commit (if you haven't committed at all yet, this is just all of the files in your project folder) and is color coded based on what I did to them.

The blue M means that I have modified an already existing (i.e. one that has already been committed through Git) file, and the yellow ? means that these are new files. If there was a red D next to any of the files, that would mean that I deleted a file that had previously been committed. There are a lot of buttons here (Diff, Commit, Pull, etc.) but you can ignore them and just click the Commit button when ready to make a commit. Doing so will open up a new window that has all the functionality of these various buttons in an easier (in my opinion) format.

This window (shown in the following image) is where you can review the changes and write up a brief note about what you did. The window is a bit overwhelming so we'll take it in pieces. First let's start by examining how the list of files in the top-left is related to the big box on the bottom with text highlighted in red and green. The list of files is identical to that in the Git tab - it's just a list of files that have changed (including new files and deleted files) since the last commit. When you click one, it'll show you the changes made to this file relative to the most recent version on Git (note that while this will show changes on R files and some other types of files, not all are available to be viewed - though that won't affect Git working at all - so it may just show a blank part of the window instead). The section that was removed is highlighted in red, and the replacement is highlighted in green. Unfortunately, it shows changes on entire lines so if you only change a small part of a line, you will have to read closely to see the difference. You can look through this to figure out exactly what you changed - both which files were changed and what was changed in each file.

Now let's walk through the process of actually committing and pushing your changes to GitHub. In real terms, this is basically uploading a new version of the files to GitHub, with brief documentation of what changed. At this point all we need to do is tell RStudio which files we want to commit, write a brief message explaining the changes, and submit it.

First, we select which files to commit by clicking the checkbox to the very left on the top left panel. In the image above, they are all unchecked as I haven't selected any yet. You can click each file's box or click the Stage button near the top once you have a file (or files) highlighted to stage it. Once it's staged the checkbox will now have a check in it. Staging a file just means that you want to commit this file. If you want to commit all of the files, you can do Control+A (or Command+A for Mac users) to select all of the files and then click Stage.

Now you're ready to document the **overall** changes that you're committing, not the changes for each individual file. You do so in the "Commit message" box on the right. Again, here it is blank but you would write a short description of the changes. There is no hard rule that it must be short, but the general convention is that each commit is relatively small and thus the description of the message can be short. You generally want no less than a short sentence and no more than a paragraph, though of course this depends on your unique circumstances. As you first start out, I think over-describing your work is best as you get a feel to what to do. Now click the Commit button.

It will make a popup window showing all the changes that it made. The "create mode ..." stuff is saying that these files are new files that Git hasn't seen before. You can close this popup.

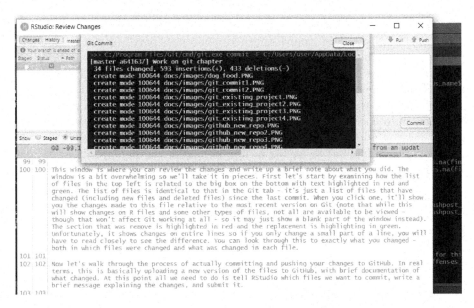

You have now completed your first commit using Git through RStudio. The files aren't on GitHub just yet though. Now right above the list of files is text that says "Your branch is ahead of 'origin/master' by 1 commit." This means that your version of the project is ahead of (since you made changes to the project that you just committed) the version on GitHub. To send it to GitHub you just need to click the Push button on the top right. In our email example, this is like clicking send after writing your draft and saving (committing) it. When you click Push it'll open up a popup, which you can close once it's done.

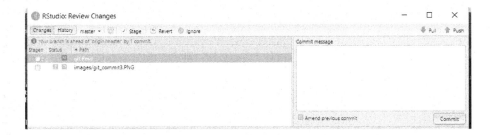

9.6   When to commit

There is no hard rule for when to make a commit, but the general convention
is to make one whenever you've finished a unique "part" of the work. For
example, if you have some data that you need to clean, graph, and run a
regression on, you'd likely commit after each part is done. One of the benefits
of using Git is that you will have a record of each version of the code that you
commit - so you want to balance between having too many records that are
very similar to each other (similar to saving a new version of a paper draft
every time you add a sentence) and too few so you lose a lot of work if you
need to go back (similar to saving a new version of the paper only every 10
pages of writing).

9.7   Other resources

For an excellent overview of using Git and GitHub with R, please see this
chapter[11] of Hadley Wickham and Jenny Bryan's book *R Packages.*[12] For a
short and very accessible book on this topic, please see Jenny Bryan and Jim
Hester's excellent *Happy Git and GitHub for the useR.*[13]

---

[11] https://r-pkgs.org/git.html
[12] https://r-pkgs.org/
[13] https://happyGitwithr.com/

Part III

Clean

10

Subsetting: Making big things small

For this chapter you'll need the following file, which is available for download here[1]: offenses_known_yearly_1960_2020.rds.

Subsetting data is a way to take a large data set and reduce it to a smaller one that is better suited for answering a specific question. This is useful when you have a lot of data in the data set that isn't relevant to your research - for example, if you are studying crime in Colorado and have every state in your data, you'd subset it to keep only the Colorado data. Reducing it to a smaller data set makes it easier to manage, both in understanding your data and avoiding have a huge file that could slow down R.

10.1 Select specific values

```
animals <- c("cat", "dog", "gorilla", "buffalo", "lion", "snake")
```

```
animals
# [1] "cat"     "dog"     "gorilla" "buffalo" "lion"
# [6] "snake"
```

Here we have made a vector object called *animals* with a number of different animals in it. In R, we will use square brackets [] to select specific values in that object, something called "indexing." Put a number (or numbers) in the square bracket, and it will return the value at that "index." The index is just the place number where each value is. "cat" is the first value in *animals* so it is at the first index, "dog" is the second value so it is the second index or

---

[1]https://github.com/jacobkap/crimebythenumbers/tree/master/data

index 2. "snake" is our last value and is the 6th value in *animals* so it is index 6.[2]

The syntax (how the code is written) goes

```
object[index]
```

First, we have the object and then we put the square bracket []. We need both the object and the [] for subsetting to work. Let's say we wanted to choose just the "snake" from our *animals* object. In normal language we say "I want the 6th value from *animals*." We say where we're looking and which value we want.

```
animals[6]
# [1] "snake"
```

Now let's get the third value.

```
animals[3]
# [1] "gorilla"
```

If we want multiple values, we can enter multiple numbers. If you have multiple values, you need to make a vector using c() and put the numbers inside the parentheses separated by a comma. If we wanted values 1-3, we could use c(1, 2, 3), with each number separated by a comma.

```
animals[c(1, 2, 3)]
# [1] "cat"      "dog"      "gorilla"
```

When making a vector of sequential integers, instead of writing them all out manually we can use first_number:last_number like so

```
1:3
# [1] 1 2 3
```

To use it in subsetting we can treat 1:3 as if we wrote c(1, 2, 3).

---

[2]Some languages use "zero indexing," which means the first index is index 0, the second is index 1. So in our example "cat" would be index 0. R does not do that, and the first value is index 1, the second is index 2, and so on.

```
animals[1:3]
# [1] "cat"      "dog"      "gorilla"
```

The order we enter the numbers determines the order of the values it returns. Let's get the third index, the fourth index, and the first index, in that order.

```
animals[c(3, 4, 1)]
# [1] "gorilla" "buffalo" "cat"
```

Putting a negative number inside the [] will return all values **except** for that index, essentially deleting it. Let's remove "cat" from *animals*. Since it is the 1st item in *animals*, we can remove it like this

```
animals[-1]
# [1] "dog"      "gorilla" "buffalo" "lion"      "snake"
```

Now let's remove multiple values, the first 3.

```
animals[-c(1, 2, 3)]
# [1] "buffalo" "lion"      "snake"
```

When using the `first_number:last_number` notation, we need to put it in parentheses if we want to turn it negative. If we don't, it will just think that the first value is a negative number, and give every integer from that first value to the last value.

```
-1:3
# [1] -1  0  1  2  3
```

Putting it in parentheses will create the integers first and then turn them all negative.

```
animals[-(1:3)]
# [1] "buffalo" "lion"      "snake"
```

Earlier I said we can remove values with using a negative number and that index will be removed from the object. For example, `animals[-1]` prints every value in *animals* except for the first value.

```
animals[-1]
# [1] "dog"     "gorilla" "buffalo" "lion"    "snake"
```

However, it doesn't actually remove anything from *animals*. Let's print *animals* and see which values it returns.

```
animals
# [1] "cat"     "dog"     "gorilla" "buffalo" "lion"
# [6] "snake"
```

Now the first value, "cats," is back. Why? To make changes in R you need to tell R very explicitly that you are making the change. If you don't save the result of your code (by assigning an object to it), R will run that code and simply print the results in the Console panel without making any changes.

This is an important point that a lot of students struggle with. R doesn't know when you want to save (in this context I am referring to creating or updating an object that is entirely in R, not saving a file to your computer) a value or update an object. If x is an object with a value of 2, and you write `x + 2`, it would print out 4 because $2 + 2 = 4$. But that won't change the value of x. x will remain as 2 until you explicitly tell R to change its value. If you want to update x you need to run `x <- somevalue` or `x = somevalue`, where "somevalue" is whatever you want to change x to.

So to return to our *animals* example, if we wanted to delete the first value and keep it removed, we'd need to write `animals <- animals[-1]`. Which is essentially making a new object, also called *animals* (to avoid having many, slightly different objects that are hard to keep track of we'll reuse the name) with the same values as the original *animals* except this time excluding the first value, "cats."

10.2   Logical values and operations

We also frequently want to conditionally select certain values. Earlier we selected values by indexing specific numbers, but that requires us to know exactly which values we want. We can conditionally select values by having some conditional statement (e.g. "this value is lower than the number 100") and keeping only values where that condition is true.

First, we will discuss conditionals abstractly and then we will use a real example using data from the FBI to make a data set tailored to answer a specific question.

We can use these TRUE and FALSE (in R true and false must be spelled all in capital letters and without quotes. For the book section on logical values, please see Section 3.1) values to index, and it will return every element which we say is TRUE.

```
animals[c(TRUE, TRUE, FALSE, FALSE, FALSE, FALSE)]
# [1] "cat" "dog"
```

This is the basis of conditional subsetting. If we have a large data set and only want a small chunk based on some condition (e.g. data for certain states, data for a certain time period, data with at least a certain population) we need to make a conditional statement that returns TRUE if it matches what we want and FALSE if it doesn't. There are a number of different ways to make conditional statements. First let's go through some special characters involved and then show examples of each one.

For each case you are asking: does the thing on the left of the conditional statement return TRUE or FALSE compared to the thing on the right.

- == Equals (compared to a single value)
- %in% Equals (one value match out of multiple comparisons)
- != Does not equal
- < Less than
- > Greater than
- <= Less than or equal to
- >= Greater than or equal to

Since many conditionals involve numbers (especially in criminology), let's make a new object called *numbers* with the numbers 1-10.

```
numbers <- 1:10
```

10.2.1 Matching a single value

The conditional == asks if the thing on the left equals the thing on the right. Note that it uses two equal signs. If we used only one equal sign it would

assign the thing on the left the value of the thing on the right (as if we did
`<-`).

```
2 == 2
# [1] TRUE
```

This gives TRUE as we know that 2 does equal 2. If we change either value, it
would give us FALSE.

```
2 == 3
# [1] FALSE
```

And it works when we have multiple numbers on the left side, such as our
object called *numbers*. This returns TRUE only for the value in *numbers* that
is 2. For all other values it returns FALSE.

```
numbers == 2
#  [1] FALSE  TRUE FALSE FALSE FALSE FALSE FALSE FALSE FALSE
# [10] FALSE
```

This also works with characters such as the animals in the object we made
earlier. "gorilla" is the third animal in our object, so if we check `animals ==`
`"gorilla"` we expect the third value to be TRUE and all others to be FALSE. Make
sure that the match is spelled correctly (including capitalization) and is in
quotes.

```
animals == "gorilla"
# [1] FALSE FALSE  TRUE FALSE FALSE FALSE
```

The `==` only works when there is one thing on the right-hand side. In criminology we often want to know if there is a match for multiple things - is the
crime one of the following crimes..., did the crime happen in one of these
months..., is the victim a member of these demographic groups...? So we
need a way to check if a value is one of many values.

10.2.2   Matching multiple values

The R operator `%in%` asks each value on the left whether or not it is a member
of the set on the right. It asks, is the single value on the left-hand side (even

when there are multiple values such as our *animals* object, it goes through them one at a time) a match with any of the values on the right-hand side? It only has to match with one of the right-hand side values to be a match.

```
2 %in% c(1, 2, 3)
# [1] TRUE
```

For our *animals* object, if we check if they are in the vector c("cat", "dog", "gorilla"), now all three of those animals will return TRUE.

```
animals %in% c("cat", "dog", "gorilla")
# [1]  TRUE  TRUE  TRUE FALSE FALSE FALSE
```

10.2.3   Does not match

Sometimes it is easier to ask what is not a match. For example, if you wanted to get every month except January, instead of writing the other 11 months, you just ask for any month that does not equal "January".

We can use !=, which means "not equal". When we wanted an exact match, we used ==, if we want a not match, we can use != (this time it is only a single equals sign).

```
2 != 3
# [1] TRUE
```

```
"cat" != "gorilla"
# [1] TRUE
```

Note that for matching multiple values with %in%, we cannot write !%in% but have to put the ! before the values on the left.

```
!animals %in% c("cat", "dog", "gorilla")
# [1] FALSE FALSE FALSE  TRUE  TRUE  TRUE
```

10.2.4   Greater than or less than

We can use R to compare values using greater than or less than symbols. We can also express "greater than or equal to" or "less than or equal to."

```
6 > 5
# [1] TRUE
```

```
6 < 5
# [1] FALSE
```

```
6 >= 5
# [1] TRUE
```

```
5 <= 5
# [1] TRUE
```

When used on our object *numbers* it will return 10 values (since *numbers* is 10 elements long) with a TRUE if the condition is true for the element and FALSE otherwise. Let's run numbers > 3. We expect the first 3 values to be FALSE as 1, 2, and 3 are not larger than 3.

```
numbers > 3
#  [1] FALSE FALSE FALSE  TRUE  TRUE  TRUE  TRUE  TRUE  TRUE
# [10]  TRUE
```

10.2.5   Combining conditional statements - or, and

In many cases when you are subsetting you will want to subset based on more than one condition. These "conditional statements" can be tricky for new R users since you need to remember both what conditions you need *and* the R code to write it. For a simple introduction to combining conditional statements, we'll first start with the dog food instructions for my new puppy Peanut.

Here, the instructions indicate how much food to feed your dog each day. Then instructions are broken down into dog age **and** expected size (in pounds or kilograms), and the intersection of these tells you how much food to feed your dog. Even once you figure out how much to feed the dog, there's another conditional statement to figure out whether you feed them twice a day or three times a day.

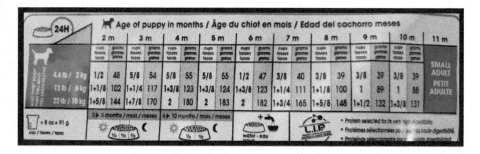

This food chart is basically a conditional statement matrix where you match the conditions on the left side with those on the top to figure out how much to feed your dog.[3]

---

[3]If you encounter some conditional statements that confuse you - which will be more common as you combine many statements together - I encourage you to make a matrix like

So if we wanted to figure out how much to feed a dog that is three months old and will be 4.4 pounds, we'd use the first row on the left (which says 4.4 pounds/2.2 kilograms) and the second column (which says three months old). When the dog gets to be four-months-old we'd keep the same row but now move one column to the right. In normal English you'd say that the dog is four months old and their expected size is 4.4 pounds (2 kg). The language when talking about (and writing code for) a conditional statement in programming is a bit more formal where every condition is spoken as a yes or no question. Here we ask is the dog four months old **and** is the expected weight 4.4 pounds? If both are true, then we give the dog the amount of food shown for those conditions. If only one is true, then the whole thing is wrong - we wouldn't want to underfeed or overfeed our dog. In this example, a four month old dog can eat between 5/8th of a cup of food and 2 cups depending on their expected size. So having only one condition be true isn't enough.

Can you see any issue with this conditional statement matrix? It doesn't cover the all possible choices for age and weight combinations. In fact, it is really quite narrow in what it does cover. For example, it covers two- and three-months, but not any age in between. We can assume that a dog that is 2.5 months old would eat the average of two and three month meal amounts, but wouldn't know for sure. When making your own statements please consider what conditions you are checking for - and, importantly, what you're leaving out.

For a real data example, let's say you have crime data from every state between 1960 and 2020. Your research question is "did Colorado's marijuana legalization affect crime in the state?" In that case you want only data from Colorado. Since legalization began in January 2014, you wouldn't need every year, only years some period of time before and after legalization to be able to measure its effect. So you would need to subset based on the state and the year.

To make conditional statements with multiple conditions we use | for "or" and & for "and".

```
Condition 1 | Condition 2
```

```
2 == 3 | 2 > 1
# [1] TRUE
```

As it sounds, when using | as long as at least one condition is true (we can include as many conditions as we like) it will return TRUE.

---

this yourself. Even if it isn't that complicated, I think it's easier to see it written down than to try to keep all of the possible conditions in your head.

```
Condition 1 & Condition 2
```

```
2 == 3 & 2 > 1
# [1] FALSE
```

For &, all of the conditions must be true. If even one condition is not true it will return FALSE.

10.3 Subsetting a data.frame

Earlier we were using a simple vector. In this book - and in your own work - you will usually work on an entire data set. These generally come in the form called a "data.frame," which you can imagine as being like an Excel file with multiple rows and columns. Section 3.3.2 covers data.frames in more detail.

Let's load in data from the Uniform Crime Report (UCR), an FBI data set that we'll work on in a later lesson. This data has crime data every year from 1960-2020 and for nearly every agency in the country.

```
ucr <- readRDS("data/offenses_known_yearly_1960_2020.rds")
```

Let's peek at the first 6 rows and 6 columns using the square bracket notation [] for data.frames, which we'll explain more below.

```
ucr[1:6, 1:6]
#        ori       ori9 agency_name   state state_abb year
# 1 AK00101 AK0010100   anchorage  alaska        AK 2020
# 2 AK00101 AK0010100   anchorage  alaska        AK 2019
# 3 AK00101 AK0010100   anchorage  alaska        AK 2018
# 4 AK00101 AK0010100   anchorage  alaska        AK 2017
# 5 AK00101 AK0010100   anchorage  alaska        AK 2016
# 6 AK00101 AK0010100   anchorage  alaska        AK 2015
```

The first 6 rows appear to be agency identification info for Anchorage, Alaska, from 2015-2020. For good measure let's check how many rows and columns are in this data. This will give us some guidance on subsetting, which we'll see below. nrow() gives us the number of rows and ncol() gives us the number of columns.

```
nrow(ucr)
# [1] 1032307
```

```
ncol(ucr)
# [1] 222
```

This is a large file with 223 columns and over a million rows. Normally we wouldn't want to print out the names of all 223 columns, but let's do so here as we want to know the variables available to subset. We can use `names()` to see the name of every column in a data.frame. Inside the parentheses we put the data.frame name (without quotes).

```
names(ucr)
#    [1] "ori"
#    [2] "ori9"
#    [3] "agency_name"
#    [4] "state"
#    [5] "state_abb"
#    [6] "year"
#    [7] "number_of_months_missing"
#    [8] "last_month_reported"
#    [9] "arson_number_of_months_missing"
#   [10] "arson_last_month_reported"
#   [11] "fips_state_code"
#   [12] "fips_county_code"
#   [13] "fips_state_county_code"
#   [14] "fips_place_code"
#   [15] "agency_type"
#   [16] "crosswalk_agency_name"
#   [17] "census_name"
#   [18] "longitude"
#   [19] "latitude"
#   [20] "address_name"
#   [21] "address_street_line_1"
#   [22] "address_street_line_2"
#   [23] "address_city"
#   [24] "address_state"
#   [25] "address_zip_code"
#   [26] "population_group"
#   [27] "population_1"
#   [28] "population_1_county"
```

```
#  [29] "population_2"
#  [30] "population_2_county"
#  [31] "population_3"
#  [32] "population_3_county"
#  [33] "population"
#  [34] "country_division"
#  [35] "juvenile_age"
#  [36] "core_city_indication"
#  [37] "fbi_field_office"
#  [38] "followup_indication"
#  [39] "zip_code"
#  [40] "month_included_in"
#  [41] "covered_by_ori"
#  [42] "agency_count"
#  [43] "special_mailing_address"
#  [44] "first_line_of_mailing_address"
#  [45] "second_line_of_mailing_address"
#  [46] "third_line_of_mailing_address"
#  [47] "fourth_line_of_mailing_address"
#  [48] "officers_killed_by_felony"
#  [49] "officers_killed_by_accident"
#  [50] "officers_assaulted"
#  [51] "actual_murder"
#  [52] "actual_manslaughter"
#  [53] "actual_rape_total"
#  [54] "actual_rape_by_force"
#  [55] "actual_rape_attempted"
#  [56] "actual_robbery_total"
#  [57] "actual_robbery_with_a_gun"
#  [58] "actual_robbery_with_a_knife"
#  [59] "actual_robbery_other_weapon"
#  [60] "actual_robbery_unarmed"
#  [61] "actual_assault_total"
#  [62] "actual_assault_with_a_gun"
#  [63] "actual_assault_with_a_knife"
#  [64] "actual_assault_other_weapon"
#  [65] "actual_assault_unarmed"
#  [66] "actual_assault_simple"
#  [67] "actual_burg_total"
#  [68] "actual_burg_force_entry"
#  [69] "actual_burg_nonforce_entry"
#  [70] "actual_burg_attempted"
#  [71] "actual_theft_total"
#  [72] "actual_mtr_veh_theft_total"
```

```
#  [73] "actual_mtr_veh_theft_car"
#  [74] "actual_mtr_veh_theft_truck"
#  [75] "actual_mtr_veh_theft_other"
#  [76] "actual_all_crimes"
#  [77] "actual_assault_aggravated"
#  [78] "actual_index_violent"
#  [79] "actual_index_property"
#  [80] "actual_index_total"
#  [81] "actual_arson_single_occupancy"
#  [82] "actual_arson_other_residential"
#  [83] "actual_arson_storage"
#  [84] "actual_arson_industrial"
#  [85] "actual_arson_other_commercial"
#  [86] "actual_arson_community_public"
#  [87] "actual_arson_all_oth_structures"
#  [88] "actual_arson_total_structures"
#  [89] "actual_arson_motor_vehicles"
#  [90] "actual_arson_other_mobile"
#  [91] "actual_arson_total_mobile"
#  [92] "actual_arson_all_other"
#  [93] "actual_arson_grand_total"
#  [94] "tot_clr_murder"
#  [95] "tot_clr_manslaughter"
#  [96] "tot_clr_rape_total"
#  [97] "tot_clr_rape_by_force"
#  [98] "tot_clr_rape_attempted"
#  [99] "tot_clr_robbery_total"
# [100] "tot_clr_robbery_with_a_gun"
# [101] "tot_clr_robbery_with_a_knife"
# [102] "tot_clr_robbery_other_weapon"
# [103] "tot_clr_robbery_unarmed"
# [104] "tot_clr_assault_total"
# [105] "tot_clr_assault_with_a_gun"
# [106] "tot_clr_assault_with_a_knife"
# [107] "tot_clr_assault_other_weapon"
# [108] "tot_clr_assault_unarmed"
# [109] "tot_clr_assault_simple"
# [110] "tot_clr_burg_total"
# [111] "tot_clr_burg_force_entry"
# [112] "tot_clr_burg_nonforce_entry"
# [113] "tot_clr_burg_attempted"
# [114] "tot_clr_theft_total"
# [115] "tot_clr_mtr_veh_theft_total"
# [116] "tot_clr_mtr_veh_theft_car"
```

```
# [117] "tot_clr_mtr_veh_theft_truck"
# [118] "tot_clr_mtr_veh_theft_other"
# [119] "tot_clr_all_crimes"
# [120] "tot_clr_assault_aggravated"
# [121] "tot_clr_index_violent"
# [122] "tot_clr_index_property"
# [123] "tot_clr_index_total"
# [124] "tot_clr_arson_single_occupancy"
# [125] "tot_clr_arson_other_residential"
# [126] "tot_clr_arson_storage"
# [127] "tot_clr_arson_industrial"
# [128] "tot_clr_arson_other_commercial"
# [129] "tot_clr_arson_community_public"
# [130] "tot_clr_arson_all_oth_structures"
# [131] "tot_clr_arson_total_structures"
# [132] "tot_clr_arson_motor_vehicles"
# [133] "tot_clr_arson_other_mobile"
# [134] "tot_clr_arson_total_mobile"
# [135] "tot_clr_arson_all_other"
# [136] "tot_clr_arson_grand_total"
# [137] "clr_18_murder"
# [138] "clr_18_manslaughter"
# [139] "clr_18_rape_total"
# [140] "clr_18_rape_by_force"
# [141] "clr_18_rape_attempted"
# [142] "clr_18_robbery_total"
# [143] "clr_18_robbery_with_a_gun"
# [144] "clr_18_robbery_with_a_knife"
# [145] "clr_18_robbery_other_weapon"
# [146] "clr_18_robbery_unarmed"
# [147] "clr_18_assault_total"
# [148] "clr_18_assault_with_a_gun"
# [149] "clr_18_assault_with_a_knife"
# [150] "clr_18_assault_other_weapon"
# [151] "clr_18_assault_unarmed"
# [152] "clr_18_assault_simple"
# [153] "clr_18_burg_total"
# [154] "clr_18_burg_force_entry"
# [155] "clr_18_burg_nonforce_entry"
# [156] "clr_18_burg_attempted"
# [157] "clr_18_theft_total"
# [158] "clr_18_mtr_veh_theft_total"
# [159] "clr_18_mtr_veh_theft_car"
# [160] "clr_18_mtr_veh_theft_truck"
```

```
# [161] "clr_18_mtr_veh_theft_other"
# [162] "clr_18_all_crimes"
# [163] "clr_18_assault_aggravated"
# [164] "clr_18_index_violent"
# [165] "clr_18_index_property"
# [166] "clr_18_index_total"
# [167] "clr_18_arson_single_occupancy"
# [168] "clr_18_arson_other_residential"
# [169] "clr_18_arson_storage"
# [170] "clr_18_arson_industrial"
# [171] "clr_18_arson_other_commercial"
# [172] "clr_18_arson_community_public"
# [173] "clr_18_arson_all_oth_structures"
# [174] "clr_18_arson_total_structures"
# [175] "clr_18_arson_motor_vehicles"
# [176] "clr_18_arson_other_mobile"
# [177] "clr_18_arson_total_mobile"
# [178] "clr_18_arson_all_other"
# [179] "clr_18_arson_grand_total"
# [180] "unfound_murder"
# [181] "unfound_manslaughter"
# [182] "unfound_rape_total"
# [183] "unfound_rape_by_force"
# [184] "unfound_rape_attempted"
# [185] "unfound_robbery_total"
# [186] "unfound_robbery_with_a_gun"
# [187] "unfound_robbery_with_a_knife"
# [188] "unfound_robbery_other_weapon"
# [189] "unfound_robbery_unarmed"
# [190] "unfound_assault_total"
# [191] "unfound_assault_with_a_gun"
# [192] "unfound_assault_with_a_knife"
# [193] "unfound_assault_other_weapon"
# [194] "unfound_assault_unarmed"
# [195] "unfound_assault_simple"
# [196] "unfound_burg_total"
# [197] "unfound_burg_force_entry"
# [198] "unfound_burg_nonforce_entry"
# [199] "unfound_burg_attempted"
# [200] "unfound_theft_total"
# [201] "unfound_mtr_veh_theft_total"
# [202] "unfound_mtr_veh_theft_car"
# [203] "unfound_mtr_veh_theft_truck"
# [204] "unfound_mtr_veh_theft_other"
```

```
# [205] "unfound_all_crimes"
# [206] "unfound_assault_aggravated"
# [207] "unfound_index_violent"
# [208] "unfound_index_property"
# [209] "unfound_index_total"
# [210] "unfound_arson_single_occupancy"
# [211] "unfound_arson_other_residential"
# [212] "unfound_arson_storage"
# [213] "unfound_arson_industrial"
# [214] "unfound_arson_other_commercial"
# [215] "unfound_arson_community_public"
# [216] "unfound_arson_all_oth_structures"
# [217] "unfound_arson_total_structures"
# [218] "unfound_arson_motor_vehicles"
# [219] "unfound_arson_other_mobile"
# [220] "unfound_arson_total_mobile"
# [221] "unfound_arson_all_other"
# [222] "unfound_arson_grand_total"
```

Now let's discuss how to subset this data into a smaller data set to answer a specific question. Let's subset the data to answer our above question of "did Colorado's marijuana legalization affect crime in the state?" Like mentioned above, we need data just from Colorado and just for years around the legalization year - we can do 2011-2017 for simplicity.

We also don't need all 223 columns in the current data. Let's say we're only interested in whether murder changes. We'd need the column called *actual_murder*, the *state* column (as a check to make sure we subset only Colorado), the *year* column, the *population* column, the *ori* column, and the *agency_name* column (a real analysis would likely grab geographic variables too to see if changes depended on location, but here we're just using it as an example). The last two columns - *ori* and *agency_name* - aren't strictly necessary but would be useful for checking if an agency's values are reasonable (e.g. see if that agency had a sudden huge spike or decline in reported crimes) when checking for outliers, a step we won't do here.

Before explaining how to subset from a data.frame, let's write pseudocode (essentially a description of what we are going to do that is readable to people but isn't real code) for our subset.

We want

- Only rows where the state equals Colorado
- Only rows where the year is 2011-2017
- Only the following columns: *actual_murder*, *state*, *year*, *population*, *ori*, *agency_name*

10.3.1   Select specific columns

The way to select a specific column in R is called the dollar sign notation.

data$column

We write the data name followed by a $ and then the column name. Make sure there are no spaces, quotation marks, or misspellings (or capitalization issues). Just the data$column exactly as it is spelled. Since we are referring to data already read into R, there should not be any quotes for either the data or the column name.

We can do this for the column *agency_name* in our UCR data. If we wrote this in the console it would print out every single row in the column. Because this data is large (over a million rows), I am going to wrap this in head() so it only displays the first 6 rows of the column rather than printing the entire column.

```
head(ucr$agency_name)
# [1] "anchorage" "anchorage" "anchorage" "anchorage"
# [5] "anchorage" "anchorage"
```

They're all the same name because Anchorage Police reported many times and are in the data set multiple times. Let's look at the column *actual_murder*, which shows the annual number of murders in that agency.

```
head(ucr$actual_murder)
# [1] 18 32 26 27 28 26
```

One hint is to write out the data set name in the console and hit the Tab key. Wait a couple of seconds and a popup will appear listing every column in the data set. You can scroll through this and then hit enter to select that column.

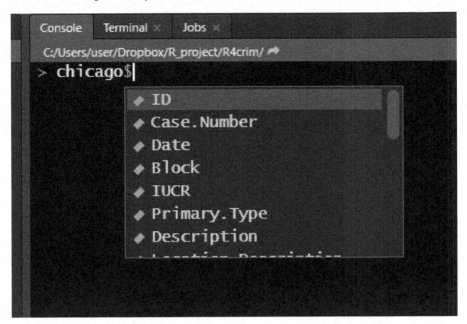

10.3.2 Select specific rows

In the earlier examples, we used square bracket notation `[]` and just put a number or several numbers in the `[]`. When dealing with data.frames, however, you need an extra step to tell R which columns to keep. The syntax in the square bracket is

`[row, column]`

We start the square bracket by saying which row we want. Now, since we also have to consider the columns, we need to tell it the number or name (in a vector using `c()` if more than one name and putting column names in quotes) of the column or columns we want.

The exception to this is when we use the dollar sign notation to select a single column. In that case we don't need a comma (and indeed it will give us an error!). Let's see a few examples and then explain why this works the way it does.

```
ucr[1, 1]
# [1] "AK00101"
```

If we input multiple numbers, we can get multiple rows and columns.

```
ucr[1:6, 1:6]
#         ori       ori9 agency_name   state state_abb year
# 1 AK00101 AK0010100   anchorage alaska          AK 2020
# 2 AK00101 AK0010100   anchorage alaska          AK 2019
# 3 AK00101 AK0010100   anchorage alaska          AK 2018
# 4 AK00101 AK0010100   anchorage alaska          AK 2017
# 5 AK00101 AK0010100   anchorage alaska          AK 2016
# 6 AK00101 AK0010100   anchorage alaska          AK 2015
```

The column section also accepts a vector of the names of the columns. These names must be spelled correctly and in quotes.

```
ucr[1:6, c("ori", "year")]
#        ori year
# 1 AK00101 2020
# 2 AK00101 2019
# 3 AK00101 2018
# 4 AK00101 2017
# 5 AK00101 2016
# 6 AK00101 2015
```

In cases where we want every row or every column, we just don't put a number. By default, R will return every row/column if you don't specify which ones you want. However, you will still need to include the comma.

Here is every column in the first row. Again, for real work we'd likely not do this as it will print out hundreds of rows to the console.

```
ucr[1, ]
#        ori       ori9 agency_name   state state_abb year
# 1 AK00101 AK0010100   anchorage alaska          AK 2020
#   number_of_months_missing last_month_reported
# 1                        0            december
#   arson_number_of_months_missing arson_last_month_reported
# 1                              0                  december
#   fips_state_code fips_county_code fips_state_county_code
# 1              02              020                  02020
#   fips_place_code          agency_type
# 1           03000 local police department
#        crosswalk_agency_name              census_name
```

```
# 1 anchorage police department anchorage municipality
#      longitude latitude                    address_name
# 1 -149.284329 61.17425 anchorage police department
#     address_street_line_1 address_street_line_2 address_city
# 1          4501 elmore rd                 <NA>    anchorage
#     address_state address_zip_code          population_group
# 1            ak            99507 city 250,000 thru 499,999
#     population_1 population_1_county population_2
# 1        286388                    0            0
#     population_2_county population_3 population_3_county
# 1                   NA            0                  NA
#     population country_division juvenile_age
# 1     286388          pacific           NA
#     core_city_indication fbi_field_office
# 1       core city of msa             3030
#           followup_indication zip_code month_included_in
# 1 do not send a follow-up     99507                    0
#     covered_by_ori agency_count      special_mailing_address
# 1           <NA>             1 not a special mailing address
#     first_line_of_mailing_address
# 1                4501 elmore rd
#     second_line_of_mailing_address
# 1                          <NA>
#     third_line_of_mailing_address
# 1                          <NA>
#     fourth_line_of_mailing_address officers_killed_by_felony
# 1                          <NA>                           0
#     officers_killed_by_accident officers_assaulted
# 1                           0                464
#     actual_murder actual_manslaughter actual_rape_total
# 1            18                   0              558
#     actual_rape_by_force actual_rape_attempted
# 1                534                  24
#     actual_robbery_total actual_robbery_with_a_gun
# 1                558                      124
#     actual_robbery_with_a_knife actual_robbery_other_weapon
# 1                          65                          82
#     actual_robbery_unarmed actual_assault_total
# 1                   287                5777
#     actual_assault_with_a_gun actual_assault_with_a_knife
# 1                512                          377
#     actual_assault_other_weapon actual_assault_unarmed
# 1                840                    609
#     actual_assault_simple actual_burg_total
```

```
# 1                    3439                1444
#    actual_burg_force_entry actual_burg_nonforce_entry
# 1                    900                       453
#    actual_burg_attempted actual_theft_total
# 1                     91               7279
#    actual_mtr_veh_theft_total actual_mtr_veh_theft_car
# 1                    1149                      807
#    actual_mtr_veh_theft_truck actual_mtr_veh_theft_other
# 1                    278                        64
#    actual_all_crimes actual_assault_aggravated
# 1                  16856                     2338
#    actual_index_violent actual_index_property
# 1                   3472                      9945
#    actual_index_total actual_arson_single_occupancy
# 1                  13417                         6
#    actual_arson_other_residential actual_arson_storage
# 1                     16                         1
#    actual_arson_industrial actual_arson_other_commercial
# 1                      0                        10
#    actual_arson_community_public
# 1                      7
#    actual_arson_all_oth_structures
# 1                      0
#    actual_arson_total_structures actual_arson_motor_vehicles
# 1                     30                        17
#    actual_arson_other_mobile actual_arson_total_mobile
# 1                      0                        17
#    actual_arson_all_other actual_arson_grand_total
# 1                      0                       73
#    tot_clr_murder tot_clr_manslaughter tot_clr_rape_total
# 1                 15                0                46
#    tot_clr_rape_by_force tot_clr_rape_attempted
# 1                     41                5
#    tot_clr_robbery_total tot_clr_robbery_with_a_gun
# 1                    207                       30
#    tot_clr_robbery_with_a_knife tot_clr_robbery_other_weapon
# 1                     33                        27
#    tot_clr_robbery_unarmed tot_clr_assault_total
# 1                    117                     3407
#    tot_clr_assault_with_a_gun tot_clr_assault_with_a_knife
# 1                    223                       281
#    tot_clr_assault_other_weapon tot_clr_assault_unarmed
# 1                    511                       428
#    tot_clr_assault_simple tot_clr_burg_total
```

```
# 1                        1964              237
#    tot_clr_burg_force_entry tot_clr_burg_nonforce_entry
# 1                        115                        118
#    tot_clr_burg_attempted tot_clr_theft_total
# 1                        4              865
#    tot_clr_mtr_veh_theft_total tot_clr_mtr_veh_theft_car
# 1                        197                        153
#    tot_clr_mtr_veh_theft_truck tot_clr_mtr_veh_theft_other
# 1                        39                        5
#    tot_clr_all_crimes tot_clr_assault_aggravated
# 1                5001                      1443
#    tot_clr_index_violent tot_clr_index_property
# 1                1711                      1326
#    tot_clr_index_total tot_clr_arson_single_occupancy
# 1                3037                      2
#    tot_clr_arson_other_residential tot_clr_arson_storage
# 1                        8                        0
#    tot_clr_arson_industrial tot_clr_arson_other_commercial
# 1                        0                        5
#    tot_clr_arson_community_public
# 1                        3
#    tot_clr_arson_all_oth_structures
# 1                        0
#    tot_clr_arson_total_structures
# 1                        13
#    tot_clr_arson_motor_vehicles tot_clr_arson_other_mobile
# 1                        5                        0
#    tot_clr_arson_total_mobile tot_clr_arson_all_other
# 1                        5                        0
#    tot_clr_arson_grand_total clr_18_murder
# 1                        27                        0
#    clr_18_manslaughter clr_18_rape_total
# 1                        0                        11
#    clr_18_rape_by_force clr_18_rape_attempted
# 1                        11                        0
#    clr_18_robbery_total clr_18_robbery_with_a_gun
# 1                        5                        2
#    clr_18_robbery_with_a_knife clr_18_robbery_other_weapon
# 1                        0                        1
#    clr_18_robbery_unarmed clr_18_assault_total
# 1                        2              228
#    clr_18_assault_with_a_gun clr_18_assault_with_a_knife
# 1                        18                        13
#    clr_18_assault_other_weapon clr_18_assault_unarmed
```

```
# 1                      25                    12
#   clr_18_assault_simple clr_18_burg_total
# 1                     160                     4
#   clr_18_burg_force_entry clr_18_burg_nonforce_entry
# 1                       4                     0
#   clr_18_burg_attempted clr_18_theft_total
# 1                       0                    36
#   clr_18_mtr_veh_theft_total clr_18_mtr_veh_theft_car
# 1                       9                     8
#   clr_18_mtr_veh_theft_truck clr_18_mtr_veh_theft_other
# 1                       1                     0
#   clr_18_all_crimes clr_18_assault_aggravated
# 1                     295                    68
#   clr_18_index_violent clr_18_index_property
# 1                      84                    51
#   clr_18_index_total clr_18_arson_single_occupancy
# 1                     135                     0
#   clr_18_arson_other_residential clr_18_arson_storage
# 1                       0                     0
#   clr_18_arson_industrial clr_18_arson_other_commercial
# 1                       0                     0
#   clr_18_arson_community_public
# 1                       0
#   clr_18_arson_all_oth_structures
# 1                       0
#   clr_18_arson_total_structures clr_18_arson_motor_vehicles
# 1                       0                     0
#   clr_18_arson_other_mobile clr_18_arson_total_mobile
# 1                       0                     0
#   clr_18_arson_all_other clr_18_arson_grand_total
# 1                       0                     2
#   unfound_murder unfound_manslaughter unfound_rape_total
# 1                       4                     0                 1
#   unfound_rape_by_force unfound_rape_attempted
# 1                       1                     0
#   unfound_robbery_total unfound_robbery_with_a_gun
# 1                       0                     0
#   unfound_robbery_with_a_knife unfound_robbery_other_weapon
# 1                       0                     0
#   unfound_robbery_unarmed unfound_assault_total
# 1                       0                     0
#   unfound_assault_with_a_gun unfound_assault_with_a_knife
# 1                       0                     0
#   unfound_assault_other_weapon unfound_assault_unarmed
```

```
# 1                           0                              0
#    unfound_assault_simple unfound_burg_total
# 1                           0                              4
#    unfound_burg_force_entry unfound_burg_nonforce_entry
# 1                           2                              1
#    unfound_burg_attempted unfound_theft_total
# 1                           1                             43
#    unfound_mtr_veh_theft_total unfound_mtr_veh_theft_car
# 1                          37                             22
#    unfound_mtr_veh_theft_truck unfound_mtr_veh_theft_other
# 1                          15                              0
#    unfound_all_crimes unfound_assault_aggravated
# 1                          89                              0
#    unfound_index_violent unfound_index_property
# 1                           5                             84
#    unfound_index_total unfound_arson_single_occupancy
# 1                          89                              0
#    unfound_arson_other_residential unfound_arson_storage
# 1                           0                              0
#    unfound_arson_industrial unfound_arson_other_commercial
# 1                           0                              0
#    unfound_arson_community_public
# 1                           0
#    unfound_arson_all_oth_structures
# 1                           0
#    unfound_arson_total_structures
# 1                           0
#    unfound_arson_motor_vehicles unfound_arson_other_mobile
# 1                           0                              0
#    unfound_arson_total_mobile unfound_arson_all_other
# 1                           0                              0
#    unfound_arson_grand_total
# 1                           0
```

Since there are 223 columns in our data, normally we'd want to avoid printing out all of them. And in most cases, we would save the output of subsets to a new object to be used later rather than just printing the output in the console.

What happens if we forget the comma? If we put in numbers for both rows and columns but don't include a comma between them it will have an error.

```
ucr[1 1]
# Error: <text>:1:7: unexpected numeric constant
```

```
# 1: ucr[1 1
#              ^
```

If we only put in a single number and no comma, it will return the column that matches that number. Here we have number 1 and it will return the first column. We'll wrap it in `head()` so it doesn't print out a million rows.

```
head(ucr[1])
#        ori
# 1 AK00101
# 2 AK00101
# 3 AK00101
# 4 AK00101
# 5 AK00101
# 6 AK00101
```

Since R thinks you are requesting a column, and we only have 223 columns in the data, asking for any number above 223 will return an error.

```
head(ucr[1000])
# Error in `[.data.frame`(ucr, 1000): undefined columns selected
```

If you already specify a column using dollar sign notation $, you do not need to indicate any column in the square brackets`[]`. All you need to do is say which row or rows you want.

```
ucr$agency_name[15]
# [1] "anchorage"
```

10.3.3   Subset Colorado data

Now we have the tools to subset our UCR data to just be Colorado from 2011-2017. There are three conditional statements we need to make, two for rows and one for columns.

- Only rows where the state equals Colorado
- Only rows where the year is 2011-2017

- Only the following columns: actual_murder, state, year, population, ori, agency_name

We could use the & operator to say rows must meet condition 1 and condition 2. Since this is an intro lesson, we will do them as two separate conditional statements. For the first step we want to get all rows in the data where the state equals "colorado" (in this data all state names are lowercase). And at this point we want to keep all columns in the data. So let's make a new object called *colorado* to save the result of this subset.

Remember that we want to put the object to the left of the [] (and touching the []) to make sure it returns the data. Just having the conditional statement will only return TRUE or FALSE values. Since we want all columns, we don't need to put anything after the comma (but we must include the comma!).

```
colorado <- ucr[ucr$state == "colorado", ]
```

Now we want to get all the rows where the year is 2011-2017. Since we want to check if the year is one of the years 2011-2017, we will use %in% and put the years in a vector 2011:2017. This time our primary data set is *colorado*, not *ucr* since *colorado* has already subsetted to just the state we want. This is how subsetting generally works. You take a large data set, subset it to a smaller one and continue to subset the smaller one to only the data you want.

```
colorado <- colorado[colorado$year %in% 2011:2017, ]
```

Finally we want the columns stated above and to keep every row in the current data. Since the format is [row, column] in this case we keep the "row" part blank to indicate that we want every row.

```
colorado <- colorado[, c(
  "actual_murder",
  "state",
  "year",
  "population",
  "ori",
  "agency_name"
)]
```

We can do a quick check using the unique() function. The unique() function prints all the unique values in a category, such as a column. We will use it

on the *state* and *year* columns to make sure only the values that we want are
present.

```
unique(colorado$state)
# [1] "colorado"
```

```
unique(colorado$year)
# [1] 2017 2016 2015 2014 2013 2012 2011
```

The only state is Colorado and the only years are 2011-2017 so our subset
worked! This data shows the number of murders in each agency. We want to
look at state trends so in Section 11.3 we will sum up all the murders per year
and see if marijuana legalization affected it.

10.3.3.1   Subsetting using `dplyr`

Above, we did subsetting through what's called the "base R" method. "Base
R" just means that we use functions that are built into R and don't use any
packages. A very popular alternative way to do most of the work done in
this chapter is to use the `dplyr` package. `dplyr` is a very useful package to
handle data and includes functions that let us subset data, select only certain
columns, and aggregate the data. For the package's website, which covers all
of the features in this package, please see here.[4]

`dplyr` is part of what is called the "tidyverse," which is a collection of R packages written by mostly the same people that include lots of functions that are
useful for working with the kind of data we use in this book. We'll cover many
of the tidyverse packages in this book. There's nothing special about a package
being a "tidyverse" package; they operate exactly the same as other packages.
I just mention it because it is a very popular set of packages, and people will
often talk about "tidyverse" approaches to R meaning using these packages.
So it's good to know the terminology. To look at the full list of tidyverse
packages, their website here[5] is an excellent overview of them.

In a lot of ways the functions we'll use from `dplyr` are simpler and easier to use
than what we wrote earlier in this chapter. In fact, a lot of people learn only
`dplyr` functions and do not learn (or at least do not spend much time on) base
R. For the rest of this book we'll use base R and tidyverse functions alongside
each other. I do this for two reasons. First, it's important to understand how

---

[4]https://dplyr.tidyverse.org/
[5]https://dplyr.tidyverse.org/

R works and using base R is the best way to learn. This is a programming-for-a-purpose book, not a pure programming book, so the focus isn't on knowing all the ins and outs of R. However, I think it is still important to have some understanding of how R works and tidyverse functions tend to obfuscate that.

In most cases this obfuscation is a good thing as it lets you focus on working with the data instead of thinking about how R works (and this is one of the tidyverse authors' motivations behind their work). In some cases, however, you'll encounter issues with either the code or your data where its important to understand how R works. In these (luckily relatively) rare cases, base R tends to be more useful in solving these problems than the tidyverse.

The second reason is that base R functions are incredibly stable. Most haven't changed since R was first created in the early 1990s. The benefit is that code you write using base R functions will work for a very long time. Using packages outside of base R (all packages, not just tidyverse packages) always carries the risk that a new version of the package will change the behavior of a function, or remove that function entirely. Thankfully this is quite rare as package developers often take care to ensure that old features remain available even as they update their package. But it is always a risk, and for programming for research we want to try to make our code as reproducible as possible, which means trying to ensure that functions we use will keep working in the future. That said, please don't avoid packages too much out of fear of this issue. Packages in R are enormously useful, and we'll use many of them throughout this book.

We'll cover two functions from `dplyr` here, and we'll also cover a couple more in the next chapter. For now, we'll look only at `filter()` and `select()`. The `filter()` function is how `dplyr` does subsetting. It takes a conditional statement and "filters" the data to only return rows where that conditional statement is true. You can include multiple conditional statements in the parentheses of `filter()` and it'll return only rows where all of the statements are true. The `select()` function does roughly that with columns where we can input a conditional statement about the name of the column (e.g. columns ending in "rate") and it'll return only those columns. `select()` also lets you choose columns just by putting the name of the column(s) in the parentheses and that's all we'll be using it for here.

Let's first copy back some of the code we used earlier when we used base R to subset Colorado data from the UCR data set.

```
colorado <- ucr[ucr$state == "colorado", ]
colorado <- colorado[colorado$year %in% 2011:2017, ]
colorado <- colorado[, c(
  "actual_murder",
  "state",
```

```
  "year",
  "population",
  "ori",
  "agency_name"
)]
```

We have two conditional statements - keep only rows where state is Colorado and where years are between 2011 and 2017 (including 2017) - and then we kept only a small number of columns.

We'll do this one step at a time using the `dplyr` functions. For `filter()` we first include the name of our data.frame, which in this case starts as "ucr" and then becomes "colorado" as we make a new object during the first line of code, and then we include our conditional statement. Using base R, we have to say which data.frame we used every time we included a column. Using `filter()` we don't need to do this. `filter()` is smart enough to select the column from the data.frame we input.

For our first filter we can write `filter(ucr, state == "colorado")` and we will save the resulting object into a data set called "colorado" like we did above. To use any `dplyr` functions we first need to install that package and then tell R we want to use it through the `library()` function.

```
install.packages("dplyr")
```

```
library(dplyr)
colorado <- filter(ucr, state == "colorado")
```

Now we can do our second conditional statement where we keep only years 2011 through 2017.

```
colorado <- filter(ucr, year %in% 2011:2017)
```

If we wanted to, we could combine these lines of code into a single line by including both conditional statements into a single `filter()` function by just including a comma after the first statement.

```
colorado <- filter(ucr, state == "colorado", year %in% 2011:2017)
```

We follow similar syntax for `select()` by starting with the name of the data set and then the name of every column you want to keep. Unlike in base R we don't need to put the columns in a vector or to put the names in quotes (though you can put the names in quotes if you'd like). The order you put the column names in is also the order it will arrange them, so this function can be used to reorder your columns.

```
colorado <- select(
  colorado, actual_murder, state,
  year, population, ori, agency_name
)
```

If we run the same checks on unique states and years as we did after our base R code, we'll get the same results. This shows that our `dplyr` code did the same thing as our base R code.

```
unique(colorado$state)
# [1] "colorado"
```

```
unique(colorado$year)
# [1] 2017 2016 2015 2014 2013 2012 2011
```

11

Exploratory data analysis

For this chapter you'll need the following files, which are available for download here[1]: ucr2017.rda and offenses_known_yearly_1960_2020.rds.

When you first start working on new data it is important to spend some time getting familiar with the data. This includes understanding how many rows and columns it has, what each row means (is each row an offender? a victim? crime in a city over a day/month/year?, etc.), and what columns it has. **Basically you want to know if the data is capable of answering the question you are asking.**

While not a comprehensive list, the following is a good start for exploratory data analysis of new data sets.

- What are the units (what does each row represent?)?
- What variables are available?
- What time period does it cover?
- Are there outliers? How many?
- Are there missing values? How many?

For the first part of this lesson we will use a data set of FBI Uniform Crime Reporting (UCR) data for 2017. This data includes every agency that reported their data for all 12 months of the year. In this part of the chapter we will look at some summary statistics for the variables we are interested in and make some basic graphs to visualize the data.

First, we need to load the data. Make sure your working directory is set to the folder where the data is.

```
load("data/ucr2017.rda")
```

The function head() will print out the first 6 rows of every column in the data. Since we only have 9 columns, we will use this function. Be careful when you have many columns (100+) as printing all of them out makes it difficult to read.

---

[1] https://github.com/jacobkap/crimebythenumbers/tree/master/data

```
head(ucr2017)
#        ori year agency_name  state population actual_murder
# 1 AK00101 2017   anchorage alaska     296188            27
# 2 AK00102 2017   fairbanks alaska      32937            10
# 3 AK00103 2017      juneau alaska      32344             1
# 4 AK00104 2017   ketchikan alaska       8230             1
# 5 AK00105 2017      kodiak alaska       6198             0
# 6 AK00106 2017        nome alaska       3829             0
#   actual_rape_total actual_robbery_total
# 1               391                  778
# 2                24                   40
# 3                50                   46
# 4                19                    0
# 5                15                    4
# 6                 7                    0
#   actual_assault_aggravated
# 1                      2368
# 2                       131
# 3                       206
# 4                        14
# 5                        41
# 6                        52
```

From these results it appears that each row is a single agency's annual data
for 2017, and the columns show the number of crimes for four crime categories
included.

Finally, we can run names() to print out every column name. We can already
see every name from head(), but this is useful when we have many columns
and don't want to use head().

```
names(ucr2017)
# [1] "ori"                        "year"
# [3] "agency_name"                "state"
# [5] "population"                 "actual_murder"
# [7] "actual_rape_total"          "actual_robbery_total"
# [9] "actual_assault_aggravated"
```

11.1   Summary and Table

An important function in understanding the data you have is `summary()` which, as discussed in Section 2.5, provides summary statistics on the numeric columns you have. Let's take a look at the results before seeing how to do something similar for categorical columns.

```
summary(ucr2017)
#       ori                  year         agency_name
#  Length:15764        Min.   :2017    Length:15764
#  Class :character    1st Qu.:2017    Class :character
#  Mode  :character    Median :2017    Mode  :character
#                      Mean   :2017
#                      3rd Qu.:2017
#                      Max.   :2017
#       state              population      actual_murder
#  Length:15764        Min.   :       0   Min.   :  0.000
#  Class :character    1st Qu.:     914   1st Qu.:  0.000
#  Mode  :character    Median :    4460   Median :  0.000
#                      Mean   :   19872   Mean   :  1.069
#                      3rd Qu.:   15390   3rd Qu.:  0.000
#                      Max.   :8616333    Max.   :653.000
#  actual_rape_total   actual_robbery_total
#  Min.   :  -2.000    Min.   :   -1.00
#  1st Qu.:   0.000    1st Qu.:    0.00
#  Median :   1.000    Median :    0.00
#  Mean   :   8.262    Mean   :   19.85
#  3rd Qu.:   5.000    3rd Qu.:    4.00
#  Max.   :2455.000    Max.   :13995.00
#  actual_assault_aggravated
#  Min.   :   -1.00
#  1st Qu.:    1.00
#  Median :    5.00
#  Mean   :   49.98
#  3rd Qu.:   21.00
#  Max.   :29771.00
```

The `table()` function returns every unique value in a category **and** how often that value appears. Unlike `summary()` we can't just put the entire data set into the (), we need to specify a single column. To specify a column you use the dollar sign notation, which is `data$column`. For most functions we use to examine the data as a whole, such as `head()`, you can do the same for a specific column.

```
head(ucr2017$agency_name)
# [1] "anchorage" "fairbanks" "juneau"    "ketchikan"
# [5] "kodiak"    "nome"
```

There are only two columns in our data with categorical values that we can use - *year* and *state*, so let's use `table()` on both of them. The columns *ori* and *agency_name* are also categorical but as each row of data has a unique ORI and name, running `table()` on those columns would not be helpful.

```
table(ucr2017$year)
#
#  2017
# 15764
```

We can see that every year in our data is 2017, as expected based on the data name. *year* is a numerical column so why can we use `table()` on it? R doesn't differentiate between numbers and characters when seeing how often each value appears. If we ran `table()` on the column "actual_murder" it would tell us how many times each unique value in the column appeared in the data. That wouldn't be very useful as we don't really care how many times an agency has, for example, 7 murders. As numeric variables often have many more unique values than character variables, it also leads to many values being printed, making it harder to understand. For columns where the number of categories is important to us, such as years, states, neighborhoods, we should use `table()`.

```
table(ucr2017$state)
#
#              alabama                alaska
#                  305                    32
#              arizona              arkansas
#                  107                   273
#           california              colorado
#                  732                   213
#          connecticut              delaware
#                  107                    63
# district of columbia               florida
#                    3                   603
#              georgia                  guam
#                  522                     1
#               hawaii                 idaho
```

```
#                        4                      95
#                 illinois                 indiana
#                      696                     247
#                     iowa                  kansas
#                      216                     309
#                 kentucky               louisiana
#                      352                     192
#                    maine                maryland
#                      135                     152
#            massachusetts                michigan
#                      346                     625
#                minnesota             mississippi
#                      397                      71
#                 missouri                 montana
#                      580                     108
#                 nebraska                  nevada
#                      225                      59
#            new hampshire              new jersey
#                      176                     576
#               new mexico                new york
#                      116                     532
#           north carolina            north dakota
#                      310                     108
#                     ohio                oklahoma
#                      532                     409
#                   oregon            pennsylvania
#                      172                    1473
#             rhode island          south carolina
#                       49                     427
#             south dakota               tennessee
#                       92                     466
#                    texas                    utah
#                      999                     125
#                  vermont                virginia
#                       85                     407
#               washington           west virginia
#                      250                     200
#                wisconsin                 wyoming
#                      433                      57
```

This shows us how many times each state is present in the data. States with a larger population tend to appear more often; this makes sense as those states have more agencies to report. Right now the results are in alphabetical order, but when knowing how frequently something appears, we usually want it ordered by frequency. We can use the sort() function to order the results from table(). Just put the entire table() function inside of the () in sort().

```
sort(table(ucr2017$state))
#
#              guam district of columbia
#                 1                      3
#              hawaii                  alaska
#                 4                      32
#          rhode island               wyoming
#                49                      57
#              nevada                 delaware
#                59                      63
#          mississippi                vermont
#                71                      85
#          south dakota                idaho
#                92                      95
#              arizona              connecticut
#               107                     107
#              montana             north dakota
#               108                     108
#            new mexico                 utah
#               116                     125
#               maine                 maryland
#               135                     152
#              oregon             new hampshire
#               172                     176
#             louisiana            west virginia
#               192                     200
#             colorado                 iowa
#               213                     216
#             nebraska                indiana
#               225                     247
#            washington              arkansas
#               250                     273
#             alabama                 kansas
#               305                     309
#          north carolina          massachusetts
#               310                     346
#             kentucky                minnesota
#               352                     397
#             virginia                oklahoma
#               407                     409
#          south carolina            wisconsin
```

```
#              427              433
#         tennessee          georgia
#              466              522
#          new york             ohio
#              532              532
#        new jersey         missouri
#              576              580
#           florida         michigan
#              603              625
#          illinois       california
#              696              732
#             texas     pennsylvania
#              999             1473
```

And if we want to sort it in decreasing order of frequency, we can use the
parameter `decreasing` in `sort()` and set it to TRUE. A parameter is just an
option used in an R function to change the way the function is used or what
output it gives. Almost all functions have these parameters, and they are useful
if you don't want to use the default setting in the function. This parameter,
`decreasing`, changes the `sort()` output to print from largest to smallest. By
default this parameter is set to FALSE, and here we say it is equal to TRUE.

```
sort(table(ucr2017$state), decreasing = TRUE)
#
#      pennsylvania             texas
#              1473              999
#        california          illinois
#               732              696
#          michigan           florida
#               625              603
#          missouri        new jersey
#               580              576
#          new york             ohio
#               532              532
#           georgia        tennessee
#               522              466
#         wisconsin    south carolina
#               433              427
#          oklahoma          virginia
#               409              407
#         minnesota          kentucky
#               397              352
#     massachusetts    north carolina
```

```
#                      346                    310
#                   kansas                alabama
#                      309                    305
#                 arkansas             washington
#                      273                    250
#                  indiana               nebraska
#                      247                    225
#                     iowa               colorado
#                      216                    213
#           west virginia              louisiana
#                      200                    192
#           new hampshire                 oregon
#                      176                    172
#                 maryland                  maine
#                      152                    135
#                     utah             new mexico
#                      125                    116
#                  montana           north dakota
#                      108                    108
#                  arizona            connecticut
#                      107                    107
#                    idaho           south dakota
#                       95                     92
#                  vermont            mississippi
#                       85                     71
#                 delaware                 nevada
#                       63                     59
#                  wyoming           rhode island
#                       57                     49
#                   alaska                 hawaii
#                       32                      4
# district of columbia                    guam
#                        3                      1
```

11.2   Graphing

We often want to make quick plots of our data to get a visual understanding of the data. We will learn a different - and in my opinion a superior - way to make graphs in Chapters 14 and 15, but for now let's use the function plot().

The plot() function is built into R so we don't need to use any packages for it.

Let's make a few scatterplots showing the relationship between two variables. With plot() the syntax (how you write the code) is plot(x_axis_variable, y_axis_variable). So all we need to do is give it the variable for the x- and y-axis. Each dot will represent a single agency (a single row in our data).

```
plot(
  ucr2017$actual_murder,
  ucr2017$actual_robbery_total
)
```

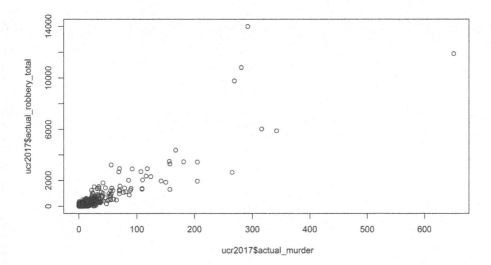

Above we are telling R to plot the number of murders on the x-axis and the number of robberies on the y-axis. This shows the relationship between a city's number of murders and number of robberies. We can see that there is a relationship where more murders is correlated with more robberies. However, there are a huge number of agencies in the bottom-left corner that have very few murders or robberies. This makes sense as - as we see in the summary() above - most agencies are small, with the median population under 5,000 people.

To try to avoid that clump of small agencies at the bottom, let's make a new data set of only agencies with a population over 1 million. We will use the filter() function from the dplyr package that was introduced in Chapter 10. For filter(), we need to first include our data set name, which is ucr2017, and then say our conditional statement. Our conditional statement is that rows in the "population" column have a value of over 1 million. For the dplyr functions we don't put our column name in quotes.

And we'll assign our results to a new object called ucr2017_big_cities Since we're using the `dplyr` package we need to tell R that we want to use it by using `library(dplyr)`.

```
library(dplyr)
ucr2017_big_cities <- filter(ucr2017, population > 1000000)
```

Now we have 18 agencies with a population of over 1 million people.

Now we can do the same graph as above but using this new data set.

```
plot(
    ucr2017_big_cities$actual_murder,
    ucr2017_big_cities$actual_robbery_total
)
```

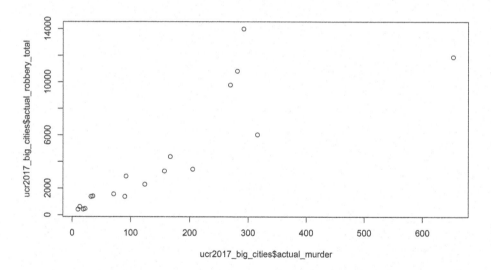

The problem is somewhat solved. There is still a small clumping of agencies with few robberies or murders, but the issue is much better. And interestingly the trend is similar with this small subset of data as with all agencies included.

To make our graph look better, we can add labels for the axes and a title (there are many options for changing the appearance of this graph, we will just use these three).

- xlab - X-axis label
- ylab - Y-axis label
- main - Graph title

Like all parameters, we add them in the () of `plot()` and separate each parameter by a comma. Since we are adding text to write in the plot, all of these parameter inputs must be in quotes.

```
plot(ucr2017_big_cities$actual_murder,
  ucr2017_big_cities$actual_robbery_total,
  xlab = "Murders",
  ylab = "Robberies",
  main = "Relationship between murder and robbery"
)
```

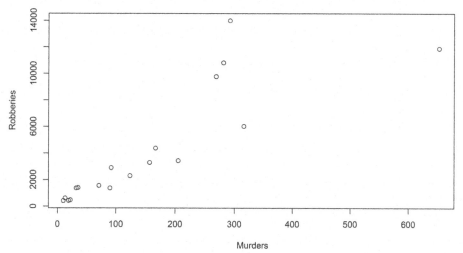

11.3 Aggregating (summaries of groups)

Right now we have the number of crimes in each agency. For many policy analyses we'd be looking at the effect on the state as a whole, rather than at the agency-level. If we wanted to do this in our data, we would need to aggregate up to the state level. Aggregating data means that we group values at some higher level than they currently are (e.g. from agency to state, from day to month, from city street to city neighborhood) and then do some mathematical operation of our choosing (in our case usually sum) to that group.

In Section 10.3.3 we started to see if marijuana legalization affected murder in Colorado. We subsetted the data to only include agencies in Colorado from 2011-2017. Now we can continue to answer the question by aggregating to the state-level to see the total number of murders per year.

Let's think about how our data are and how we would (theoretically, before we write any code) find that out.

Our data has a single row for each agency, and we have a column indicating the year the agency reported. So how would we find out how many murders happened in Colorado for each year? Well, first we take all the agencies in 2011 (the first year we're looking at) and add up the murders for all agencies that reported that year. Then take all the rows in 2012 and add up their murders. And so on for all the years.

To do this in R, we'll be using two new functions from the `dplyr` package: `group_by()` and `summarize()`.

These functions do the aggregation process in two steps. First we use `group_by()` to tell R which columns we want to group our data by - these are the higher level of aggregation columns so in our case will be the year of data. Then we need to sum up the number of murders each year. We do this using `summarize()`, and we'll specify in the function that we want to sum up the data, rather than use some other math operation on it like finding the average number of murders each year.

First, let's load back in the data and then repeat the subsetting code we did in Section 10.3.3 to keep only data for Colorado from 2011 through 2017. We'll also include the "actual_robbery_total" column that we excluded in Section 10.3.3 so we can see how easy it is to aggregate multiple columns at once using this method.

```
ucr <- readRDS("data/offenses_known_yearly_1960_2020.rds")
colorado <- filter(
  ucr, state == "colorado",
  year %in% 2011:2017
)
colorado <- select(
  colorado, actual_murder, actual_robbery_total,
  state, year, population, ori, agency_name
)
```

First we must group the data by using the `group_by()` function. Here we're just grouping the data by year, but we could group it by multiple columns if we

want by adding a comma and then the next column we want. Following other
dplyr function syntax, we first input the data set name and then the column
name - neither of which need to be in quotes.

```
colorado <- group_by(colorado, year)
```

Now we can summarize the data using the summarize() function. As with other
dplyr functions the first input is the data set name. Then we choose our math
function (sum, mean, median, etc.) and just apply that function on the column
we want. So in our case we want the sum of murders so we use sum() and include
the column we want to aggregate inside of sum()'s parentheses.

```
summarize(colorado, sum(actual_murder))
# # A tibble: 7 x 2
#     year `sum(actual_murder)`
#    <dbl>                <dbl>
# 1  2011                   154
# 2  2012                   163
# 3  2013                   172
# 4  2014                   148
# 5  2015                   173
# 6  2016                   203
# 7  2017                   218
```

If we want to aggregate another column we just add a comma after our initial
column and add another math operation function and the column we want.
Here we're also using sum(), but we could use different math operations if we
want - they don't need to be the same.

```
summarize(
  colorado, sum(actual_murder),
  sum(actual_robbery_total)
)
# # A tibble: 7 x 3
#     year `sum(actual_murder)` `sum(actual_robbery_total)`
#    <dbl>                <dbl>                       <dbl>
# 1  2011                   154                        3287
# 2  2012                   163                        3369
# 3  2013                   172                        3122
# 4  2014                   148                        3021
# 5  2015                   173                        3305
```

```
# 6   2016                203                     3513
# 7   2017                218                     3811
```

We could even do different math operations on the same column and we'd get multiple columns from it. Let's add another column showing the average number of robberies as an example.

```
summarize(
  colorado, sum(actual_murder),
  sum(actual_robbery_total),
  mean(actual_robbery_total)
)
# # A tibble: 7 x 4
#     year `sum(actual_murder)` sum(actual_robbery_to~1 mean(~2
#    <dbl>               <dbl>                   <dbl>  <dbl>
# 1   2011                 154                    3287   11.2
# 2   2012                 163                    3369   11.2
# 3   2013                 172                    3122   10.3
# 4   2014                 148                    3021    9.94
# 5   2015                 173                    3305   10.9
# 6   2016                 203                    3513   11.6
# 7   2017                 218                    3811   12.5
# # ... with abbreviated variable names
# #   1: `sum(actual_robbery_total)`,
# #   2: `mean(actual_robbery_total)`
```

By default `summarize()` calls the columns it makes using what we include in the parentheses. Since we said "sum(actual_murder)", to get the sum of the murder column, it names that new column "sum(actual_murder)". Usually we'll want to name the columns ourselves. We can do this by assigning the summarized column to a name using "name =" before it. For example, we could write "murders = sum(actual_murder)" and it will name that column "murders" instead of "sum(actual_murder)". Like other things in dplyr functions, we don't need to put quotes around our new column name. We'll assign this final summarized data to an object called "colorado_agg" so we can use it to make graphs. And to be able to create crime rates per population, we'll also find the sum of the population for each year.

```
colorado_agg <- summarize(colorado,
  murders    = sum(actual_murder),
  robberies  = sum(actual_robbery_total),
```

```
  population = sum(population)
)
colorado_agg
# # A tibble: 7 x 4
#    year murders robberies population
#   <dbl>   <dbl>     <dbl>      <dbl>
# 1  2011     154      3287    5155993
# 2  2012     163      3369    5227884
# 3  2013     172      3122    5308236
# 4  2014     148      3021    5402555
# 5  2015     173      3305    5505856
# 6  2016     203      3513    5590124
# 7  2017     218      3811    5661529
```

Now we can see that the total number of murders increased over time. So can we conclude that marijuana legalization increases murder? No, all this analysis shows is that the years following marijuana legalization, murders increased in Colorado. But that can be due to many reasons other than marijuana. For a proper analysis you'd need a comparison state that is similar to Colorado prior to legalization (and that didn't legalize marijuana) and see if their murders changes following Colorado's legalization.

To control for population, we'll standardize our murder data by creating a murder rate per 100,000 people. We can do this by dividing the murder column by the population column and then multiplying by 100,000. Let's do that and assign the result into a new column called "murder_rate".

```
colorado_agg$murder_rate <- colorado_agg$murders /
   colorado_agg$population * 100000
```

If we also wanted a robbery rate we'd do the same with the robberies column.

```
colorado_agg$robbery_rate <- colorado_agg$robberies /
   colorado_agg$population * 100000
```

The `dplyr` package has a helpful function that can do this too, and allows us to do it while writing less code. The `mutate()` function lets us create or alter columns in our data. Like other `dplyr` functions we start by including our data set in the parentheses, and then we can follow standard assignment (covered in Section 2.2) though we must use = here and not <-. A benefit of using `mutate()` is that we don't have to write out our data set name each time. So we'd write `murder_rate = murders / population * 100000`. And if we wanted to

make two (or more) columns at the same time we just add a comma after our
first assignment and then do the next assignment.

```
mutate(colorado_agg,
    murder_rate  = murders / population * 100000,
    robbery_rate = robberies / population * 100000
)
# # A tibble: 7 x 6
#     year murders robberies population murder_rate robbery_r~1
#    <dbl>   <dbl>     <dbl>      <dbl>       <dbl>       <dbl>
# 1  2011     154      3287    5155993        2.99        63.8
# 2  2012     163      3369    5227884        3.12        64.4
# 3  2013     172      3122    5308236        3.24        58.8
# 4  2014     148      3021    5402555        2.74        55.9
# 5  2015     173      3305    5505856        3.14        60.0
# 6  2016     203      3513    5590124        3.63        62.8
# 7  2017     218      3811    5661529        3.85        67.3
# # ... with abbreviated variable name 1: robbery_rate
```

Now let's make a plot of this data showing the murder rate over time. With
time-series graphs we want the time variable to be on the x-axis and the
numeric variable that we are measuring to be on the y-axis.

```
plot(
    x = colorado_agg$year,
    y = colorado_agg$murder_rate
)
```

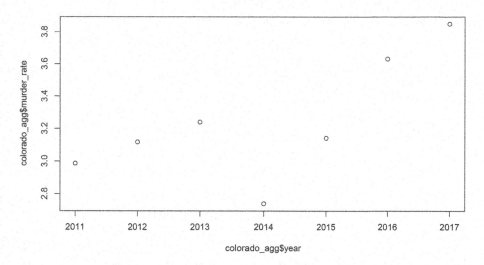

By default `plot()` makes a scatterplot. If we set the parameter `type` to "l" it will be a line plot.

```
plot(
  x = colorado_agg$year,
  y = colorado_agg$murder_rate,
  type = "l"
)
```

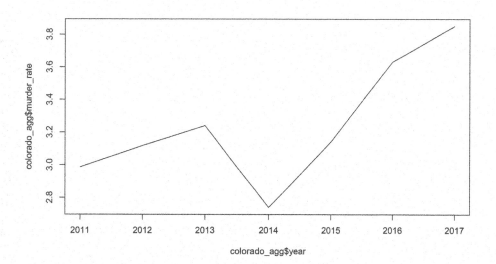

We can add some labels and a title to make this graph easier to read.

```
plot(
  x = colorado_agg$year,
  y = colorado_agg$murder_rate,
  type = "l",
  xlab = "Year",
  ylab = "Murders per 100k Population",
  main = "Murder Rate in Colorado, 2011-2017"
)
```

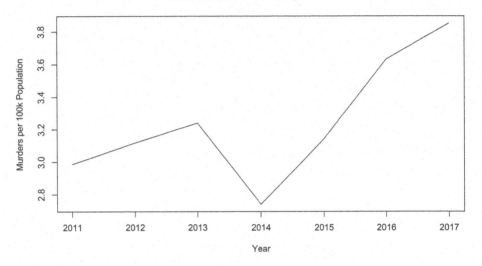

11.4   Pipes in `dplyr`

To end this chapter we'll talk about something called a pipe that is a very useful and powerful part of `dplyr`.

Think about the math equation $1 + 2 + 3 + 4$. Here we know that we add 1 and 2 together, and then add the result to 3 and then add the result of that to 4. This is much simpler to write than splitting everything up and summing each value together in a different line. In terms of R, we have so far been doing things as if we could only add two numbers together and then need a separate line to add the third (and another line to add the fourth) number.

For example, below are the two lines of code we used to subset the data to just the right state and years we wanted, and the columns we wanted. We did this in two separate lines. In our math example, we did $1 + 2$. And then found the answer, and separately did $3 + 3$. And then again found the answer and did $6 + 4$.

```
colorado <- filter(
  ucr, state == "colorado",
  year %in% 2011:2017
)
colorado <- select(
  colorado,
  actual_murder,
  actual_robbery_total,
  state,
  year,
  population,
  ori,
  agency_name
)
head(colorado)
#    actual_murder actual_robbery_total    state year
# 1              7                   80 colorado 2017
# 2             11                   93 colorado 2016
# 3              6                   68 colorado 2015
# 4              6                   58 colorado 2014
# 5              7                   44 colorado 2013
# 6              7                   55 colorado 2012
#    population      ori agency_name
# 1       99940 CO00100        adams
# 2      100526 CO00100        adams
# 3      100266 CO00100        adams
# 4       98569 CO00100        adams
# 5       97146 CO00100        adams
# 6       93542 CO00100        adams
```

With `dplyr` we actually do have a way to chain together functions; to do the programming equivalent of $1 + 2 + 3 + 4$ all at once.[2] We do this through what is called a pipe, which allows us to take the result of one function and immediately put it into another function without having to save the initial

---

[2]Pipes are technically from the `magrittr` package, but we'll just be using pipes in the context of using functions from `dplyr` or other tidyverse packages.

result or start a new line of code. To use a pipe we put the following code after the end of a function: `%>%`.

These three characters, `%>%`, are the pipe, and they must be written exactly like this. The pipe is itself actually a function, but it is a special type of function we won't go into detail about. Personally I don't think this really looks like a pipe at all, but it is called a pipe so that's the terminology I'll be using. How a pipe technically works is that it takes the output of the initial function (which is usually a tibble, which is the tidyverse's modified version of a data.frame) and puts it automatically in the first input in the next function. This won't work for all functions but nearly all functions from the tidyverse collection of packages have a data set as the first input so it will work here. The benefit is that we don't need to keep saving the output from functions or specifying which data set to include in each function.

As an example, we'll rewrite the previous code using a pipe. We start with our data.frame, which is normally the first thing we put in any `dplyr` function, and then immediately have a pipe `%>%` into a `dplyr` function, which here is `filter()`. Now we don't need to say what the data set is because it takes the last thing that was piped into the function, which in our case is the entire data.frame ucr. After our `filter()` is done we have another pipe and go into `select()`. Now `select()` will use as its first input whatever is outputted from the `filter()`. So the input to `select()` will be the subsetted data output from `filter()`. We can have as many pipes as we wish, and chain many different `dplyr` functions together, but we just use two functions here so we'll end after our `select()` function.

```
colorado <- ucr %>%
  filter(
    state == "colorado",
    year %in% 2011:2017
  ) %>%
  select(
    actual_murder,
    actual_robbery_total,
    state, year, population,
    ori, agency_name
  )
```

If we check results using `head()`, we can see that this code is exactly the same as not using pipes.

```
head(colorado)
#    actual_murder actual_robbery_total    state year
```

```
# 1                   7                80 colorado 2017
# 2                  11                93 colorado 2016
# 3                   6                68 colorado 2015
# 4                   6                58 colorado 2014
# 5                   7                44 colorado 2013
# 6                   7                55 colorado 2012
#    population      ori agency_name
# 1      99940 CO00100        adams
# 2     100526 CO00100        adams
# 3     100266 CO00100        adams
# 4      98569 CO00100        adams
# 5      97146 CO00100        adams
# 6      93542 CO00100        adams
```

The normal way to write code using pipes is to have a new line after the pipe
and after each comma in `filter()` and `select()`. This doesn't change how the
code works at all, but it is easier to read now because it has less code bunched
together in a single line.

```
colorado <- ucr %>%
  filter(
    state == "colorado",
    year %in% 2011:2017
  ) %>%
  select(
    actual_murder,
    actual_robbery_total,
    state,
    year,
    population,
    ori,
    agency_name
  )
```

12

Regular Expressions

Many word processing programs like Microsoft Word or Google Docs let you search for a pattern - usually a word or phrase - and it will show you where on the page that pattern appears. It also lets you replace that word or phrase with something new. R does the same using the function `grep()` to search for a pattern and tell you where in the data it appears, and `gsub()`, which lets you search for a pattern and then replace it with a new pattern.

* `grep()` - Find
* `gsub()` - Find and Replace

The `grep()` function lets you find a pattern in the text and it will return a number saying which element has the pattern (in a data.frame this tells you which row has a match). `gsub()` lets you input a pattern to find and a pattern to replace it with, just like Find and Replace features elsewhere. You can remember the difference because `gsub()` has the word "sub" in it, and what it does is **sub**stitute text with new text.

A useful cheat sheet on regular expressions is available here[1].

For this lesson we will use a vector of 50 crime categories. These are all of the crimes in San Francisco Police data. As we'll see, there are some issues with the crime names that we need to fix.

```
crimes <- c(
  "Arson",
  "Assault",
  "Burglary",
  "Case Closure",
  "Civil Sidewalks",
  "Courtesy Report",
  "Disorderly Conduct",
  "Drug Offense",
  "Drug Violation",
  "Embezzlement",
```

---

[1] https://www.rstudio.com/wp-content/uploads/2016/09/RegExCheatsheet.pdf

```
    "Family Offense",
    "Fire Report",
    "Forgery And Counterfeiting",
    "Fraud",
    "Gambling",
    "Homicide",
    "Human Trafficking (A), Commercial Sex Acts",
    "Human Trafficking, Commercial Sex Acts",
    "Juvenile Offenses",
    "Larceny Theft",
    "Liquor Laws",
    "Lost Property",
    "Malicious Mischief",
    "Miscellaneous Investigation",
    "Missing Person",
    "Motor Vehicle Theft",
    "Motor Vehicle Theft?",
    "Non-Criminal",
    "Offences Against The Family And Children",
    "Other",
    "Other Miscellaneous",
    "Other Offenses",
    "Prostitution",
    "Rape",
    "Recovered Vehicle",
    "Robbery",
    "Sex Offense",
    "Stolen Property",
    "Suicide",
    "Suspicious",
    "Suspicious Occ",
    "Traffic Collision",
    "Traffic Violation Arrest",
    "Vandalism",
    "Vehicle Impounded",
    "Vehicle Misplaced",
    "Warrant",
    "Weapons Carrying Etc",
    "Weapons Offence",
    "Weapons Offense"
)
```

When looking closely at these crimes it is clear that some may overlap in certain categories such as theft, and there are several duplicates with slight differences in spelling. For example the last two crimes are "Weapons Offence" and "Weapons Offense." These should be the same crime but the first one spelled "offense" wrong. And take a look at "motor vehicle theft." There are two crimes here because one of them adds a question mark at the end for some reason.

12.1   Finding patterns in text with grep()

We'll start with grep() which allows us to search a vector of data (in R, columns in a data.frame operate the same as a vector) and find where there is a match for the pattern we want to look for.

The syntax for grep() is

```
grep("pattern", data)
```

where pattern is the pattern you are searching for, such as "a" if you want to find all values with the letter a. The pattern must always be in quotes. data is a vector of strings (such as *crimes* we made above or a column in a data.frame) that you are searching in to find the pattern.

The output of this function is a number that says which element(s) in the vector the pattern was found in. If it returns, for example, the numbers 1 and 3 you know that the first and third element in your vector has the pattern - and that no other elements do. It is essentially returning the index where the conditional statement "is this pattern present" is true.

So since our data is *crimes* our grep() function will be grep("", crimes). What we put in the " " is the pattern we want to search for.

Let's start with the letter "a".

```
grep("a", crimes)
#  [1]  2  3  4  5  9 11 14 15 17 18 20 21 23 24 28 29 31 34
# [19] 42 43 44 46 47 48 49 50
```

It gives us a bunch of numbers where the letter "a" is present in that element of *crimes*. This is useful for subsetting. We can use grep() to find all values that match a pattern we want and subset to keep just those values.

```
crimes[grep("a", crimes)]
#  [1] "Assault"
#  [2] "Burglary"
#  [3] "Case Closure"
#  [4] "Civil Sidewalks"
#  [5] "Drug Violation"
#  [6] "Family Offense"
#  [7] "Fraud"
#  [8] "Gambling"
#  [9] "Human Trafficking (A), Commercial Sex Acts"
# [10] "Human Trafficking, Commercial Sex Acts"
# [11] "Larceny Theft"
# [12] "Liquor Laws"
# [13] "Malicious Mischief"
# [14] "Miscellaneous Investigation"
# [15] "Non-Criminal"
# [16] "Offences Against The Family And Children"
# [17] "Other Miscellaneous"
# [18] "Rape"
# [19] "Traffic Collision"
# [20] "Traffic Violation Arrest"
# [21] "Vandalism"
# [22] "Vehicle Misplaced"
# [23] "Warrant"
# [24] "Weapons Carrying Etc"
# [25] "Weapons Offence"
# [26] "Weapons Offense"
```

Searching for the letter "a" isn't that useful. Let's say we want to subset the data to only include theft-related crimes. From reading the list of crimes we can see there are multiple theft crimes - "Larceny Theft", "Motor Vehicle Theft", and "Motor Vehicle Theft?". We may also want to include "Stolen Property" in this search, but we'll wait until later in this lesson for how to search for multiple patterns. Since those three crimes all have the word "Theft" in the name we can search for that pattern, and it will return only those crimes.

```
grep("Theft", crimes)
# [1] 20 26 27
```

```
crimes[grep("Theft", crimes)]
# [1] "Larceny Theft"         "Motor Vehicle Theft"
# [3] "Motor Vehicle Theft?"
```

A very useful parameter in `grep()` is `value`. When we set `value` to TRUE, it will print out the actual strings that are a match rather than the element number. While this prevents us from using it to subset (since R no longer knows which rows are a match), it is an excellent tool to check if the `grep()` was successful as we can visually confirm it returns what we want. When we start to learn about special characters that make the patterns more complicated, this will be important.

```
grep("Theft", crimes, value = TRUE)
# [1] "Larceny Theft"         "Motor Vehicle Theft"
# [3] "Motor Vehicle Theft?"
```

Note that `grep()` (and `gsub()`) is case sensitive so you must capitalize properly.

```
grep("theft", crimes, value = TRUE)
# character(0)
```

Setting the parameter `ignore.case` to be TRUE makes `grep()` ignore capitalization.

```
grep("theft", crimes, value = TRUE, ignore.case = TRUE)
# [1] "Larceny Theft"         "Motor Vehicle Theft"
# [3] "Motor Vehicle Theft?"
```

If we want to find values that do *not* match with "theft", we can set the parameter `invert` to TRUE.

```
grep("theft", crimes, value = TRUE, ignore.case = TRUE, invert = TRUE)
#  [1] "Arson"
#  [2] "Assault"
#  [3] "Burglary"
#  [4] "Case Closure"
#  [5] "Civil Sidewalks"
#  [6] "Courtesy Report"
#  [7] "Disorderly Conduct"
```

```
#  [8] "Drug Offense"
#  [9] "Drug Violation"
# [10] "Embezzlement"
# [11] "Family Offense"
# [12] "Fire Report"
# [13] "Forgery And Counterfeiting"
# [14] "Fraud"
# [15] "Gambling"
# [16] "Homicide"
# [17] "Human Trafficking (A), Commercial Sex Acts"
# [18] "Human Trafficking, Commercial Sex Acts"
# [19] "Juvenile Offenses"
# [20] "Liquor Laws"
# [21] "Lost Property"
# [22] "Malicious Mischief"
# [23] "Miscellaneous Investigation"
# [24] "Missing Person"
# [25] "Non-Criminal"
# [26] "Offences Against The Family And Children"
# [27] "Other"
# [28] "Other Miscellaneous"
# [29] "Other Offenses"
# [30] "Prostitution"
# [31] "Rape"
# [32] "Recovered Vehicle"
# [33] "Robbery"
# [34] "Sex Offense"
# [35] "Stolen Property"
# [36] "Suicide"
# [37] "Suspicious"
# [38] "Suspicious Occ"
# [39] "Traffic Collision"
# [40] "Traffic Violation Arrest"
# [41] "Vandalism"
# [42] "Vehicle Impounded"
# [43] "Vehicle Misplaced"
# [44] "Warrant"
# [45] "Weapons Carrying Etc"
# [46] "Weapons Offence"
# [47] "Weapons Offense"
```

12.2   Finding and replacing patterns in text with gsub()

gsub() takes patterns and replaces them with other patterns. An important use in criminology for gsub() is to fix spelling mistakes in the text, such as the way "offense" was spelled wrong in our data. This will be a standard part of your data cleaning process and is important as a misspelled word can cause significant issues. For example if our previous example of marijuana legalization in Colorado had half of agencies misspelling the name "Colorado", aggregating the data by the state (or simply subsetting to just Colorado agencies) would give completely different results as you'd lose half your data.

gsub() is also useful when you want to take subcategories and change the value to larger categories. For example we could take any crime with the word "Theft" in it and change the whole crime name to "Theft". In our data that would take 3 subcategories of thefts and turn it into a larger category we could aggregate to. This will be useful in city-level data where you may only care about a certain type of crime but it has many subcategories that you need to aggregate.

The syntax of gsub() is similar to grep() with the addition of a pattern to replace the pattern we found.

```
gsub("find_pattern", "replace_pattern", data)
```

Let's start with a simple example of finding the letter "a" and replacing it with "z". Our data will be the word "cat".

```
gsub("a", "z", "cat")
# [1] "czt"
```

Like grep(), gsub() is case sensitive and has the parameter ignore.case to ignore capitalization.

```
gsub("A", "z", "cat")
# [1] "cat"
```

```
gsub("A", "z", "cat", ignore.case = TRUE)
# [1] "czt"
```

`gsub()` returns the same data you input but with the pattern already replaced. Above you can see that when using capital A, it returns "cat" unchanged as it never found the pattern. When `ignore.case` was set to TRUE it returned "czt" as it then matched to letter "A".

We can use `gsub()` to replace some issues in the crimes data, such as "Offense" being spelled "Offence".

```
gsub("Offence", "Offense", crimes)
#  [1] "Arson"
#  [2] "Assault"
#  [3] "Burglary"
#  [4] "Case Closure"
#  [5] "Civil Sidewalks"
#  [6] "Courtesy Report"
#  [7] "Disorderly Conduct"
#  [8] "Drug Offense"
#  [9] "Drug Violation"
# [10] "Embezzlement"
# [11] "Family Offense"
# [12] "Fire Report"
# [13] "Forgery And Counterfeiting"
# [14] "Fraud"
# [15] "Gambling"
# [16] "Homicide"
# [17] "Human Trafficking (A), Commercial Sex Acts"
# [18] "Human Trafficking, Commercial Sex Acts"
# [19] "Juvenile Offenses"
# [20] "Larceny Theft"
# [21] "Liquor Laws"
# [22] "Lost Property"
# [23] "Malicious Mischief"
# [24] "Miscellaneous Investigation"
# [25] "Missing Person"
# [26] "Motor Vehicle Theft"
# [27] "Motor Vehicle Theft?"
# [28] "Non-Criminal"
# [29] "Offenses Against The Family And Children"
# [30] "Other"
# [31] "Other Miscellaneous"
# [32] "Other Offenses"
# [33] "Prostitution"
# [34] "Rape"
# [35] "Recovered Vehicle"
```

```
# [36] "Robbery"
# [37] "Sex Offense"
# [38] "Stolen Property"
# [39] "Suicide"
# [40] "Suspicious"
# [41] "Suspicious Occ"
# [42] "Traffic Collision"
# [43] "Traffic Violation Arrest"
# [44] "Vandalism"
# [45] "Vehicle Impounded"
# [46] "Vehicle Misplaced"
# [47] "Warrant"
# [48] "Weapons Carrying Etc"
# [49] "Weapons Offense"
# [50] "Weapons Offense"
```

A useful pattern is an empty string " " which says replace whatever the find_pattern is with nothing, deleting it. Let's delete the letter "a" (lowercase only) from the data.

```
gsub("a", "", crimes)
#  [1] "Arson"
#  [2] "Assult"
#  [3] "Burglry"
#  [4] "Cse Closure"
#  [5] "Civil Sidewlks"
#  [6] "Courtesy Report"
#  [7] "Disorderly Conduct"
#  [8] "Drug Offense"
#  [9] "Drug Violtion"
# [10] "Embezzlement"
# [11] "Fmily Offense"
# [12] "Fire Report"
# [13] "Forgery And Counterfeiting"
# [14] "Frud"
# [15] "Gmbling"
# [16] "Homicide"
# [17] "Humn Trfficking (A), Commercil Sex Acts"
# [18] "Humn Trfficking, Commercil Sex Acts"
# [19] "Juvenile Offenses"
# [20] "Lrceny Theft"
# [21] "Liquor Lws"
# [22] "Lost Property"
```

```
# [23] "Mlicious Mischief"
# [24] "Miscellneous Investigtion"
# [25] "Missing Person"
# [26] "Motor Vehicle Theft"
# [27] "Motor Vehicle Theft?"
# [28] "Non-Criminl"
# [29] "Offences Aginst The Fmily And Children"
# [30] "Other"
# [31] "Other Miscellneous"
# [32] "Other Offenses"
# [33] "Prostitution"
# [34] "Rpe"
# [35] "Recovered Vehicle"
# [36] "Robbery"
# [37] "Sex Offense"
# [38] "Stolen Property"
# [39] "Suicide"
# [40] "Suspicious"
# [41] "Suspicious Occ"
# [42] "Trffic Collision"
# [43] "Trffic Violtion Arrest"
# [44] "Vndlism"
# [45] "Vehicle Impounded"
# [46] "Vehicle Misplced"
# [47] "Wrrnt"
# [48] "Wepons Crrying Etc"
# [49] "Wepons Offence"
# [50] "Wepons Offense"
```

12.3   Useful special characters

So far, we have just searched for a single character or word and expected
a return only if an exact match was found. Now we'll discuss a number of
characters called "special characters" that allow us to make more complex
`grep()` and `gsub()` pattern searches.

12.3.1   Multiple characters []

To search for multiple matches we can put the pattern we want to search for inside square brackets [] (note that we use the same square brackets for subsetting, but they operate very differently in this context). For example, we can find all the crimes that contain the letters "x", "y", or "z".

The grep() searches if any of the letters inside of the [] are present in our *crimes* vector.

```
grep("[xyz]", crimes, value = TRUE)
#  [1] "Burglary"
#  [2] "Courtesy Report"
#  [3] "Disorderly Conduct"
#  [4] "Embezzlement"
#  [5] "Family Offense"
#  [6] "Forgery And Counterfeiting"
#  [7] "Human Trafficking (A), Commercial Sex Acts"
#  [8] "Human Trafficking, Commercial Sex Acts"
#  [9] "Larceny Theft"
# [10] "Lost Property"
# [11] "Offences Against The Family And Children"
# [12] "Robbery"
# [13] "Sex Offense"
# [14] "Stolen Property"
# [15] "Weapons Carrying Etc"
```

As it searches for any letter inside of the square brackets, the order does not matter.

```
grep("[zyx]", crimes, value = TRUE)
#  [1] "Burglary"
#  [2] "Courtesy Report"
#  [3] "Disorderly Conduct"
#  [4] "Embezzlement"
#  [5] "Family Offense"
#  [6] "Forgery And Counterfeiting"
#  [7] "Human Trafficking (A), Commercial Sex Acts"
#  [8] "Human Trafficking, Commercial Sex Acts"
#  [9] "Larceny Theft"
# [10] "Lost Property"
# [11] "Offences Against The Family And Children"
# [12] "Robbery"
```

```
# [13] "Sex Offense"
# [14] "Stolen Property"
# [15] "Weapons Carrying Etc"
```

This also works for numbers though we do not have any numbers in the data.

```
grep("[01234567890]", crimes, value = TRUE)
# character(0)
```

If we wanted to search for a pattern, such as vowels, that is repeated we could put multiple [] patterns together. We will see another way to search for a repeated pattern soon.

```
grep("[aeiou][aeiou][aeiou]", crimes, value = TRUE)
# [1] "Malicious Mischief"
# [2] "Miscellaneous Investigation"
# [3] "Other Miscellaneous"
# [4] "Suspicious"
# [5] "Suspicious Occ"
```

Inside the [] we can also use the dash sign - to make intervals between certain values. For numbers, n-m means any number between n and m (inclusive). For letters, a-z means all lowercase letters and A-Z means all uppercase letters in that range (inclusive).

```
grep("[x-z]", crimes, value = TRUE)
#  [1] "Burglary"
#  [2] "Courtesy Report"
#  [3] "Disorderly Conduct"
#  [4] "Embezzlement"
#  [5] "Family Offense"
#  [6] "Forgery And Counterfeiting"
#  [7] "Human Trafficking (A), Commercial Sex Acts"
#  [8] "Human Trafficking, Commercial Sex Acts"
#  [9] "Larceny Theft"
# [10] "Lost Property"
# [11] "Offences Against The Family And Children"
# [12] "Robbery"
# [13] "Sex Offense"
```

```
# [14] "Stolen Property"
# [15] "Weapons Carrying Etc"
```

12.3.2   n-many of previous character {n}

{n} means the preceding item will be matched exactly n times.

We can use it to rewrite the above grep() to say the values in the [] should be repeated three times.

```
grep("[aeiou]{3}", crimes, value = TRUE)
# [1] "Malicious Mischief"
# [2] "Miscellaneous Investigation"
# [3] "Other Miscellaneous"
# [4] "Suspicious"
# [5] "Suspicious Occ"
```

12.3.3   n-many to m-many of previous character {n,m}

While {n} says "the previous character (or characters inside a []) must be present exactly n times", we can allow a range by using {n,m}. Here the previous character must be present between n and m times (inclusive).

We can check for values where there are 2-3 vowels in a row. Note that there cannot be a space before or after the comma.

```
grep("[aeiou]{2,3}", crimes, value = TRUE)
#  [1] "Assault"
#  [2] "Courtesy Report"
#  [3] "Drug Violation"
#  [4] "Forgery And Counterfeiting"
#  [5] "Fraud"
#  [6] "Human Trafficking (A), Commercial Sex Acts"
#  [7] "Human Trafficking, Commercial Sex Acts"
#  [8] "Liquor Laws"
#  [9] "Malicious Mischief"
# [10] "Miscellaneous Investigation"
# [11] "Offences Against The Family And Children"
# [12] "Other Miscellaneous"
# [13] "Prostitution"
```

```
# [14] "Suicide"
# [15] "Suspicious"
# [16] "Suspicious Occ"
# [17] "Traffic Collision"
# [18] "Traffic Violation Arrest"
# [19] "Vehicle Impounded"
# [20] "Weapons Carrying Etc"
# [21] "Weapons Offence"
# [22] "Weapons Offense"
```

If we wanted only crimes with exactly three vowels in a row we'd use {3,3}.

```
grep("[aeiou]{3,3}", crimes, value = TRUE)
# [1] "Malicious Mischief"
# [2] "Miscellaneous Investigation"
# [3] "Other Miscellaneous"
# [4] "Suspicious"
# [5] "Suspicious Occ"
```

If we leave n blank, such as {,m}, it says, "previous character must be present up to m times."

```
grep("[aeiou]{,3}", crimes, value = TRUE)
#  [1] "Arson"
#  [2] "Assault"
#  [3] "Burglary"
#  [4] "Case Closure"
#  [5] "Civil Sidewalks"
#  [6] "Courtesy Report"
#  [7] "Disorderly Conduct"
#  [8] "Drug Offense"
#  [9] "Drug Violation"
# [10] "Embezzlement"
# [11] "Family Offense"
# [12] "Fire Report"
# [13] "Forgery And Counterfeiting"
# [14] "Fraud"
# [15] "Gambling"
# [16] "Homicide"
# [17] "Human Trafficking (A), Commercial Sex Acts"
# [18] "Human Trafficking, Commercial Sex Acts"
```

```
# [19] "Juvenile Offenses"
# [20] "Larceny Theft"
# [21] "Liquor Laws"
# [22] "Lost Property"
# [23] "Malicious Mischief"
# [24] "Miscellaneous Investigation"
# [25] "Missing Person"
# [26] "Motor Vehicle Theft"
# [27] "Motor Vehicle Theft?"
# [28] "Non-Criminal"
# [29] "Offences Against The Family And Children"
# [30] "Other"
# [31] "Other Miscellaneous"
# [32] "Other Offenses"
# [33] "Prostitution"
# [34] "Rape"
# [35] "Recovered Vehicle"
# [36] "Robbery"
# [37] "Sex Offense"
# [38] "Stolen Property"
# [39] "Suicide"
# [40] "Suspicious"
# [41] "Suspicious Occ"
# [42] "Traffic Collision"
# [43] "Traffic Violation Arrest"
# [44] "Vandalism"
# [45] "Vehicle Impounded"
# [46] "Vehicle Misplaced"
# [47] "Warrant"
# [48] "Weapons Carrying Etc"
# [49] "Weapons Offence"
# [50] "Weapons Offense"
```

This returns every crime as "up to m times" includes zero times.

And the same works for leaving m blank, but it will be "present at least n times".

```
grep("[aeiou]{3,}", crimes, value = TRUE)
# [1] "Malicious Mischief"
# [2] "Miscellaneous Investigation"
# [3] "Other Miscellaneous"
```

```
# [4] "Suspicious"
# [5] "Suspicious Occ"
```

12.3.4   Start of string

The ∧ symbol (called a caret) signifies that what follows it is the start of the
string. We put the ∧ at the beginning of the quotes and then anything that
follows it must be the very start of the string. As an example let's search for
"Family". Our data has both the "Family Offense" crime and the "Offences
Against The Family And Children" crime (which likely are the same crime
written differently). If we use ∧ then we should only have the first one returned.

```
grep("^Family", crimes, value = TRUE)
# [1] "Family Offense"
```

12.3.5   End of string $

The dollar sign $ acts similar to the caret ∧ except that it signifies that the
value before it is the **end** of the string. We put the $ at the very end of our
search pattern and whatever character is before it is the end of the string. For
example, let's search for all crimes that end with the word "Theft".

```
grep("Theft$", crimes, value = TRUE)
# [1] "Larceny Theft"        "Motor Vehicle Theft"
```

Note that the crime "Motor Vehicle Theft?" doesn't get selected as it ends
with a question mark.

12.3.6   Anything .

The . symbol is a stand-in for any value. This is useful when you aren't
sure about every part of the pattern you are searching. It can also be used
when there are slight differences in words such as our incorrect "Offence" and
"Offense". We can replace the "c" and "s" with the ..

```
grep("Weapons Offen.e", crimes, value = TRUE)
# [1] "Weapons Offence" "Weapons Offense"
```

12.3.7 One or more of previous +

The + means that the character immediately before it is present at least one time. This is the same as writing {1,}. If we wanted to find all values with only two words, we would start with some number of letters followed by a space followed by some more letters and the string would end.

```
grep("^[A-Za-z]+ [A-Za-z]+$", crimes, value = TRUE)
#  [1] "Case Closure"
#  [2] "Civil Sidewalks"
#  [3] "Courtesy Report"
#  [4] "Disorderly Conduct"
#  [5] "Drug Offense"
#  [6] "Drug Violation"
#  [7] "Family Offense"
#  [8] "Fire Report"
#  [9] "Juvenile Offenses"
# [10] "Larceny Theft"
# [11] "Liquor Laws"
# [12] "Lost Property"
# [13] "Malicious Mischief"
# [14] "Miscellaneous Investigation"
# [15] "Missing Person"
# [16] "Other Miscellaneous"
# [17] "Other Offenses"
# [18] "Recovered Vehicle"
# [19] "Sex Offense"
# [20] "Stolen Property"
# [21] "Suspicious Occ"
# [22] "Traffic Collision"
# [23] "Vehicle Impounded"
# [24] "Vehicle Misplaced"
# [25] "Weapons Offence"
# [26] "Weapons Offense"
```

12.3.8 Zero or more of previous *

The * special character says match zero or more of the previous character and is the same as {0,}. Combining . with * is powerful when used in gsub() to delete text before or after a pattern. Let's write a pattern that searches the text for the word "Weapons" and then deletes any text after that.

Our pattern would be "Weapons.*" which is the word "Weapons" followed by anything zero or more times.

```
gsub("Weapons.*", "Weapons", crimes)
#  [1] "Arson"
#  [2] "Assault"
#  [3] "Burglary"
#  [4] "Case Closure"
#  [5] "Civil Sidewalks"
#  [6] "Courtesy Report"
#  [7] "Disorderly Conduct"
#  [8] "Drug Offense"
#  [9] "Drug Violation"
# [10] "Embezzlement"
# [11] "Family Offense"
# [12] "Fire Report"
# [13] "Forgery And Counterfeiting"
# [14] "Fraud"
# [15] "Gambling"
# [16] "Homicide"
# [17] "Human Trafficking (A), Commercial Sex Acts"
# [18] "Human Trafficking, Commercial Sex Acts"
# [19] "Juvenile Offenses"
# [20] "Larceny Theft"
# [21] "Liquor Laws"
# [22] "Lost Property"
# [23] "Malicious Mischief"
# [24] "Miscellaneous Investigation"
# [25] "Missing Person"
# [26] "Motor Vehicle Theft"
# [27] "Motor Vehicle Theft?"
# [28] "Non-Criminal"
# [29] "Offences Against The Family And Children"
# [30] "Other"
# [31] "Other Miscellaneous"
# [32] "Other Offenses"
# [33] "Prostitution"
# [34] "Rape"
# [35] "Recovered Vehicle"
# [36] "Robbery"
# [37] "Sex Offense"
# [38] "Stolen Property"
# [39] "Suicide"
# [40] "Suspicious"
# [41] "Suspicious Occ"
# [42] "Traffic Collision"
# [43] "Traffic Violation Arrest"
```

```
# [44] "Vandalism"
# [45] "Vehicle Impounded"
# [46] "Vehicle Misplaced"
# [47] "Warrant"
# [48] "Weapons"
# [49] "Weapons"
# [50] "Weapons"
```

And now our last three crimes are all the same.

12.3.9 Multiple patterns |

The vertical bar | special character allows us to check for multiple patterns. It essentially functions as "pattern A or Pattern B" with the | symbol replacing the word "or" (and making sure to not have any space between patterns.). To check our crimes for the word "Drug" or the word "Weapons", we could write "Drug|Weapon", which searches for "Drug" or "Weapons" in the text.

```
grep("Drug|Weapons", crimes, value = TRUE)
# [1] "Drug Offense"         "Drug Violation"
# [3] "Weapons Carrying Etc" "Weapons Offence"
# [5] "Weapons Offense"
```

12.3.10 Parentheses ()

Parentheses act similar to the square brackets [], where we want everything inside but with parentheses the values must be in the proper order.

```
grep("(Offense)", crimes, value = TRUE)
# [1] "Drug Offense"      "Family Offense"
# [3] "Juvenile Offenses" "Other Offenses"
# [5] "Sex Offense"       "Weapons Offense"
```

Running the above code returns the same results as if we didn't include the parentheses. The usefulness of parentheses comes when combining it with the | symbol to be able to check "(X|Y) Z"), which says, "look for either X or Y which must be followed by Z".

Running just "(Offense)" returns values for multiple types of offenses. Let's say we just care about Drug and Weapon Offenses. We can search for "Offense" normally and combine () and | to say, "search for either the word 'Drug' or the word 'Family' and they should be followed by the word 'Offense'."

```
grep("(Drug|Weapons) Offense", crimes, value = TRUE)
# [1] "Drug Offense"    "Weapons Offense"
```

12.3.11   Optional text ?

The question mark indicates that the character immediately before the ? is optional.

Let's search for the term "offens" and add a ? at the end. This says search for the pattern "offen", and we expect an exact match for that pattern. And if the letter "s" follows "offen" return that too, but it isn't required to be there.

```
grep("Offens?", crimes, value = TRUE)
# [1] "Drug Offense"
# [2] "Family Offense"
# [3] "Juvenile Offenses"
# [4] "Offences Against The Family And Children"
# [5] "Other Offenses"
# [6] "Sex Offense"
# [7] "Weapons Offence"
# [8] "Weapons Offense"
```

We can further combine it with () and | to get both spellings of Weapon Offense.

```
grep("(Drug|Weapons) Offens?", crimes, value = TRUE)
# [1] "Drug Offense"    "Weapons Offence" "Weapons Offense"
```

12.4   Changing capitalization

If you're dealing with data where the only difference is capitalization (as is common in crime data) instead of using `gsub()` to change individual values, you can use the functions `toupper()` and `tolower()` to change every letter's capitalization. These functions take as an input a vector of strings (or a column from a data.frame) and return those strings either upper or lowercase.

```
toupper(crimes)
#  [1] "ARSON"
#  [2] "ASSAULT"
#  [3] "BURGLARY"
#  [4] "CASE CLOSURE"
#  [5] "CIVIL SIDEWALKS"
#  [6] "COURTESY REPORT"
#  [7] "DISORDERLY CONDUCT"
#  [8] "DRUG OFFENSE"
#  [9] "DRUG VIOLATION"
# [10] "EMBEZZLEMENT"
# [11] "FAMILY OFFENSE"
# [12] "FIRE REPORT"
# [13] "FORGERY AND COUNTERFEITING"
# [14] "FRAUD"
# [15] "GAMBLING"
# [16] "HOMICIDE"
# [17] "HUMAN TRAFFICKING (A), COMMERCIAL SEX ACTS"
# [18] "HUMAN TRAFFICKING, COMMERCIAL SEX ACTS"
# [19] "JUVENILE OFFENSES"
# [20] "LARCENY THEFT"
# [21] "LIQUOR LAWS"
# [22] "LOST PROPERTY"
# [23] "MALICIOUS MISCHIEF"
# [24] "MISCELLANEOUS INVESTIGATION"
# [25] "MISSING PERSON"
# [26] "MOTOR VEHICLE THEFT"
# [27] "MOTOR VEHICLE THEFT?"
# [28] "NON-CRIMINAL"
# [29] "OFFENCES AGAINST THE FAMILY AND CHILDREN"
# [30] "OTHER"
# [31] "OTHER MISCELLANEOUS"
# [32] "OTHER OFFENSES"
# [33] "PROSTITUTION"
# [34] "RAPE"
# [35] "RECOVERED VEHICLE"
# [36] "ROBBERY"
# [37] "SEX OFFENSE"
# [38] "STOLEN PROPERTY"
# [39] "SUICIDE"
# [40] "SUSPICIOUS"
```

```
# [41] "SUSPICIOUS OCC"
# [42] "TRAFFIC COLLISION"
# [43] "TRAFFIC VIOLATION ARREST"
# [44] "VANDALISM"
# [45] "VEHICLE IMPOUNDED"
# [46] "VEHICLE MISPLACED"
# [47] "WARRANT"
# [48] "WEAPONS CARRYING ETC"
# [49] "WEAPONS OFFENCE"
# [50] "WEAPONS OFFENSE"
```

```
tolower(crimes)
# [1] "arson"
# [2] "assault"
# [3] "burglary"
# [4] "case closure"
# [5] "civil sidewalks"
# [6] "courtesy report"
# [7] "disorderly conduct"
# [8] "drug offense"
# [9] "drug violation"
# [10] "embezzlement"
# [11] "family offense"
# [12] "fire report"
# [13] "forgery and counterfeiting"
# [14] "fraud"
# [15] "gambling"
# [16] "homicide"
# [17] "human trafficking (a), commercial sex acts"
# [18] "human trafficking, commercial sex acts"
# [19] "juvenile offenses"
# [20] "larceny theft"
# [21] "liquor laws"
# [22] "lost property"
# [23] "malicious mischief"
# [24] "miscellaneous investigation"
# [25] "missing person"
# [26] "motor vehicle theft"
# [27] "motor vehicle theft?"
# [28] "non-criminal"
# [29] "offences against the family and children"
# [30] "other"
```

```
# [31] "other miscellaneous"
# [32] "other offenses"
# [33] "prostitution"
# [34] "rape"
# [35] "recovered vehicle"
# [36] "robbery"
# [37] "sex offense"
# [38] "stolen property"
# [39] "suicide"
# [40] "suspicious"
# [41] "suspicious occ"
# [42] "traffic collision"
# [43] "traffic violation arrest"
# [44] "vandalism"
# [45] "vehicle impounded"
# [46] "vehicle misplaced"
# [47] "warrant"
# [48] "weapons carrying etc"
# [49] "weapons offence"
# [50] "weapons offense"
```

13

Reshaping data

For this chapter you'll need the following file, which is available for download here[1]: sqf-2019.xlsx. This file was initially downloaded from the New York City Police Department's page here[2].

When you're using data for research, the end result is usually a regression or a graph (or both), and that requires your data to be in a particular format. Usually your data should have one row for each unit of analysis, and each column should have information about that unit. As an example, if you wanted to study city-level crime over time, you'd have each row be a single city in a single time period. If you looked at 10 different time periods, say 10 years, you'd have 10 rows for each city. And each column would have information about that city in that time period, such as the number of murders that occurred.

This is what is known as "long" format, as this data often has many rows and few columns. The alternative is what's known as "wide" format, where each row (in our example) is a single agency and you'd have 10 times as many columns in "long" data as there are 10 time periods. Whereas with "long" data you'd have, for example, a "murder" column and 10 rows showing the murder in each year, with "wide" data you'd have 10 murder columns showing the value for each year's murder count. While long data is more commonly used in criminology research, some statistical and graphing approaches require wide data. If this seems like an abstract concept, bear with it for a bit as we'll go over several examples in this chapter.

Ideally, our data comes in the exact format we need. In cases where it doesn't, we need to be able to convert the data from long to wide or from wide to long format. This conversion is called "reshaping". There are many approaches and packages to doing reshaping in R, which makes reshaping actually one of the most confusing things to do in R. In this chapter we'll use an approach from the package `tidyr`, which, as the name suggests, is one of the packages from the tidyverse. To use this package we need to install it using `install.packages()`.

---

[1] https://github.com/jacobkap/crimebythenumbers/tree/master/data
[2] https://www1.nyc.gov/site/nypd/stats/reports-analysis/stopfrisk.page

```
install.packages("tidyr")
```

For this chapter we'll use microdata from the New York City Police Department for every stop and frisk they conducted in 2019. The data is in an .xlsx format so we can use the readxl package to read it into R. Let's call the data "sqf" as it is the abbreviation for "stop, question, and frisk," which is the full name of the data set. Following traditional R naming conventions, we'll keep the object name lowercased.

```
library(readxl)
sqf <- read_excel("data/sqf-2019.xlsx")
```

The first thing we want to do when loading in a new data set to R is to look at it. We can do this a few different ways, including using View(), which opens up an Excel-like tab where we can scroll through the data, and through head() which prints the first 6 rows of every column to the console. The readxl package converts the data into a "tibble". which is essentially a modified data.frame. One of the modifications is that it doesn't print out every single column when we use head() to view the first 6 rows of each column. That's because tibble's don't want to print large amounts of text to the console.

Normally we'd use head() to look at the data, but as there are dozens of columns in this data, I don't want to print out the first six rows of every column to the console as it will take up a lot of space in the book. If this were real data you were working on, you would use head() or View() to look at the data. Instead, here we'll use names() to see what columns are in the data. Let's also use nrow() to see how many rows of data we have.

```
sqf <- data.frame(sqf)
names(sqf)
#  [1] "STOP_ID_ANONY"
#  [2] "STOP_FRISK_DATE"
#  [3] "STOP_FRISK_TIME"
#  [4] "YEAR2"
#  [5] "MONTH2"
#  [6] "DAY2"
#  [7] "STOP_WAS_INITIATED"
#  [8] "RECORD_STATUS_CODE"
#  [9] "ISSUING_OFFICER_RANK"
# [10] "ISSUING_OFFICER_COMMAND_CODE"
# [11] "SUPERVISING_OFFICER_RANK"
# [12] "SUPERVISING_OFFICER_COMMAND_CODE"
```

```
# [13]  "LOCATION_IN_OUT_CODE"
# [14]  "JURISDICTION_CODE"
# [15]  "JURISDICTION_DESCRIPTION"
# [16]  "OBSERVED_DURATION_MINUTES"
# [17]  "SUSPECTED_CRIME_DESCRIPTION"
# [18]  "STOP_DURATION_MINUTES"
# [19]  "OFFICER_EXPLAINED_STOP_FLAG"
# [20]  "OFFICER_NOT_EXPLAINED_STOP_DESCRIPTION"
# [21]  "OTHER_PERSON_STOPPED_FLAG"
# [22]  "SUSPECT_ARRESTED_FLAG"
# [23]  "SUSPECT_ARREST_OFFENSE"
# [24]  "SUMMONS_ISSUED_FLAG"
# [25]  "SUMMONS_OFFENSE_DESCRIPTION"
# [26]  "OFFICER_IN_UNIFORM_FLAG"
# [27]  "ID_CARD_IDENTIFIES_OFFICER_FLAG"
# [28]  "SHIELD_IDENTIFIES_OFFICER_FLAG"
# [29]  "VERBAL_IDENTIFIES_OFFICER_FLAG"
# [30]  "FRISKED_FLAG"
# [31]  "SEARCHED_FLAG"
# [32]  "ASK_FOR_CONSENT_FLG"
# [33]  "CONSENT_GIVEN_FLG"
# [34]  "OTHER_CONTRABAND_FLAG"
# [35]  "FIREARM_FLAG"
# [36]  "KNIFE_CUTTER_FLAG"
# [37]  "OTHER_WEAPON_FLAG"
# [38]  "WEAPON_FOUND_FLAG"
# [39]  "PHYSICAL_FORCE_CEW_FLAG"
# [40]  "PHYSICAL_FORCE_DRAW_POINT_FIREARM_FLAG"
# [41]  "PHYSICAL_FORCE_HANDCUFF_SUSPECT_FLAG"
# [42]  "PHYSICAL_FORCE_OC_SPRAY_USED_FLAG"
# [43]  "PHYSICAL_FORCE_OTHER_FLAG"
# [44]  "PHYSICAL_FORCE_RESTRAINT_USED_FLAG"
# [45]  "PHYSICAL_FORCE_VERBAL_INSTRUCTION_FLAG"
# [46]  "PHYSICAL_FORCE_WEAPON_IMPACT_FLAG"
# [47]  "BACKROUND_CIRCUMSTANCES_VIOLENT_CRIME_FLAG"
# [48]  "BACKROUND_CIRCUMSTANCES_SUSPECT_KNOWN_TO_CARRY_WEAPON_FLAG"
# [49]  "SUSPECTS_ACTIONS_CASING_FLAG"
# [50]  "SUSPECTS_ACTIONS_CONCEALED_POSSESSION_WEAPON_FLAG"
# [51]  "SUSPECTS_ACTIONS_DECRIPTION_FLAG"
# [52]  "SUSPECTS_ACTIONS_DRUG_TRANSACTIONS_FLAG"
# [53]  "SUSPECTS_ACTIONS_IDENTIFY_CRIME_PATTERN_FLAG"
# [54]  "SUSPECTS_ACTIONS_LOOKOUT_FLAG"
# [55]  "SUSPECTS_ACTIONS_OTHER_FLAG"
# [56]  "SUSPECTS_ACTIONS_PROXIMITY_TO_SCENE_FLAG"
```

```
# [57]  "SEARCH_BASIS_ADMISSION_FLAG"
# [58]  "SEARCH_BASIS_CONSENT_FLAG"
# [59]  "SEARCH_BASIS_HARD_OBJECT_FLAG"
# [60]  "SEARCH_BASIS_INCIDENTAL_TO_ARREST_FLAG"
# [61]  "SEARCH_BASIS_OTHER_FLAG"
# [62]  "SEARCH_BASIS_OUTLINE_FLAG"
# [63]  "DEMEANOR_CODE"
# [64]  "DEMEANOR_OF_PERSON_STOPPED"
# [65]  "SUSPECT_REPORTED_AGE"
# [66]  "SUSPECT_SEX"
# [67]  "SUSPECT_RACE_DESCRIPTION"
# [68]  "SUSPECT_HEIGHT"
# [69]  "SUSPECT_WEIGHT"
# [70]  "SUSPECT_BODY_BUILD_TYPE"
# [71]  "SUSPECT_EYE_COLOR"
# [72]  "SUSPECT_HAIR_COLOR"
# [73]  "SUSPECT_OTHER_DESCRIPTION"
# [74]  "STOP_LOCATION_PRECINCT"
# [75]  "STOP_LOCATION_SECTOR_CODE"
# [76]  "STOP_LOCATION_APARTMENT"
# [77]  "STOP_LOCATION_FULL_ADDRESS"
# [78]  "STOP_LOCATION_STREET_NAME"
# [79]  "STOP_LOCATION_X"
# [80]  "STOP_LOCATION_Y"
# [81]  "STOP_LOCATION_ZIP_CODE"
# [82]  "STOP_LOCATION_PATROL_BORO_NAME"
# [83]  "STOP_LOCATION_BORO_NAME"
nrow(sqf)
# [1] 13459
```

Each row of data is a stop and frisk, and there were 13,459 in 2019. That number may seem low to you, especially if you've read articles about how frequent stop and frisks happens in New York City. It is correct - at least correct in terms of reported stops. Stop and frisks in New York City peaked in 2011 with nearly 700,000 conducted, and then fell sharply after that to fewer than 23,000 from 2015 through 2019 (the last year available at the time of this writing).

Looking at the results of `names()`, we can see that this is a rich data set with lots of information about each stop (though many columns also have missing data, so it is not as rich as it initially appears). Each stop has, for example, the date and time of the stop, whether an arrest was made and for what, physical characteristics (race, sex, age, height, weight, clothing) of the person stopped, and the location of the stop. One important variable that's missing is a unique identifier for either the officer or the person stopped so we can see how often they are in the data. The "ISSUING_OFFICER_COMMAND_CODE" variable may be a unique ID for the officer, but it's not clear that it is - there are different officer ranks for the same command code so this appears to me to be different officers.

13.1   Reshaping a single column

Relative to most data sets in criminology, this is an enormous amount of information. So I encourage you to explore this data and practice your R skills. As this chapter focuses on reshaping, we won't do much exploring of the data. This data is in long format as each row is a different stop, and the columns are information about that specific stop. We can't convert this to wide format as there are no repeated entries in the data; each row is of a unique person stopped (at least as far as we can tell. In reality, there are likely to be many people stopped multiple times in the data). We could use another variable such as stopped persons' race or gender and reshape it using that, but that'll lead to many thousands of columns so is not a great idea.

Instead, we'll first aggregate the data and then reshape that aggregated data. We'll aggregate the data by month and by day of week and see how many people of each race were stopped at each month-day-of-week. In addition to `tidyr`, we'll use functions from `dplyr` to aggregate our data. Since we've already installed that package we just need to use `library()` and don't need to use `install.packages()` again.

We've already used the `group_by()` function in aggregating, and now we'll introduce a new function: `count()`. In our earlier aggregation code (introduced in Section 11.3), we used the function `summarize()` and summed up columns that had a numeric variable (e.g. the number of murders). `count()` works similarly by sums up categorical variables, which in our case is the race of people stopped. Using `count()` we enter in the categorical column in the parentheses, and it'll make a new column called "n", which is the sum of each category. Since we're aggregating the data let's assign the result of our aggregating to a new object called "sqf_agg".

```
library(tidyr)
library(dplyr)
sqf_agg <- sqf %>%
  group_by(
    MONTH2,
```

```
    DAY2
 ) %>%
 count(SUSPECT_RACE_DESCRIPTION)
```

Now let's look at the `head()` of the result.

```
head(sqf_agg)
# # A tibble: 6 x 4
# # Groups:   MONTH2, DAY2 [1]
#   MONTH2 DAY2   SUSPECT_RACE_DESCRIPTION        n
#   <chr>  <chr>  <chr>                       <int>
# 1 April  Friday (null)                          1
# 2 April  Friday ASIAN / PACIFIC ISLANDER        1
# 3 April  Friday BLACK                         104
# 4 April  Friday BLACK HISPANIC                 17
# 5 April  Friday WHITE                          16
# 6 April  Friday WHITE HISPANIC                 31
```

Each row is now a month-day-of-week-race combination, and we have two additional columns: first the person stopped's race and then the "n" column, which says how many people of that race were stopped in that month-day-of-week. The first race is "(null)", which means we don't know the race. We could keep these values as an "unknown" race, but for simplicity we'll just remove them using `filter()`.

To make sure we only have values we expect, we can use the `unique()` function to check that we have only months and week-days we expect, and that our race names are consistent. This is important because even though you know, for example, that there are only seven days in a week, there may be more in our data due to typos or erroneous data entry. So it is always good to check.

```
unique(sqf_agg$MONTH2)
#  [1] "April"      "August"    "December"   "February"
#  [5] "January"    "July"      "June"       "March"
#  [9] "May"        "November"  "October"    "September"
unique(sqf_agg$DAY2)
#  [1] "Friday"     "Monday"    "Saturday"   "Sunday"
#  [5] "Thursday"   "Tuesday"   "Wednesday"
unique(sqf_agg$SUSPECT_RACE_DESCRIPTION)
#  [1] "(null)"                   "ASIAN / PACIFIC ISLANDER"
#  [3] "BLACK"                    "BLACK HISPANIC"
```

```
# [5] "WHITE"                      "WHITE HISPANIC"
# [7] "AMERICAN INDIAN/ALASKAN N"
```

The months and days of the week look good. For an actual data analysis we may want to combine the Hispanic values to a single "Hispanic" value instead of splitting it between "BLACK HISPANIC" and "WHITE HISPANIC", but for this lesson we'll leave it as it is.

Now, we want to reshape the data from its current long format to a wide format. We do this using the `pivot_wider()` function from the `tidyr` package. In the function we need to input two values, the name of the column, which identifies what each value in the row means, and the column that has that value. In our case this is the "SUSPECT_RACE_DESCRIPTION" column and the "n" column. We don't need to put the column names in quotes. These function parameters are called "names_from" and "values_from" so we'll also include that in the `pivot_wider()` function. And finally before we run `pivot_wider()`, we'll first use `filter()` to remove any row where the race is "(null)". We'll assign the result of this code to an object called "sqf_agg_wide" and use `head()` to look at this result.

```
sqf_agg_wide <- sqf_agg %>%
  filter(SUSPECT_RACE_DESCRIPTION != "(null)") %>%
  pivot_wider(
    names_from = SUSPECT_RACE_DESCRIPTION,
    values_from = n
  )
head(sqf_agg_wide)
# # A tibble: 6 x 8
# # Groups:   MONTH2, DAY2 [6]
#   MONTH2 DAY2     ASIAN~1 BLACK BLACK~2 WHITE WHITE~3 AMERI~4
#   <chr>  <chr>      <int> <int>   <int> <int>   <int>   <int>
# 1 April  Friday         1   104      17    16      31      NA
# 2 April  Monday         1    92      10    32      29      NA
# 3 April  Saturd~        3   115      24    24      44      NA
# 4 April  Sunday         2    96      12    15      38      NA
# 5 April  Thursd~        2   122      10    18      38      NA
# 6 April  Tuesday        7   137      14    20      49      NA
# # ... with abbreviated variable names
# #   1: `ASIAN / PACIFIC ISLANDER`, 2: `BLACK HISPANIC`,
# #   3: `WHITE HISPANIC`, 4: `AMERICAN INDIAN/ALASKAN N`
```

Now instead of having one row be a month-day-of-week-race combination, each row is a month-day-of-week pair, and we have one column for every race in our

data. Each of these race columns tell us how many people of that race were
stopped in that month-day-of-week. This allows for really easy comparison of
things like racial differences in stops for each month-day-of-week as we just
look at different columns in the same row. We have now successfully done our
first reshaping, moving this data from long to wide format!

Now we want to reshape it again to go from wide to long. Our race columns
names just took the values from the race column, which means that we have
column names all in capital letters and with spaces and slashes in them. We
could technically use this as it is, but it's a bit trickier to use any name
with punctuation in it, and is against R column name convention, so we'll
quickly fix this. To do so we'll use the `make_clean_names()` function from the
`janitor` package that automatically makes all character inputs to the function
lowercase and replaces all punctuation with an underscore. First, we need to
install the package with `install.packages()`.

```
install.packages("janitor")
```

To use this function on the column names we'll make the input of the function
`names(sqf_agg_wide)`, which returns the names of the "sqf_agg_wide" data, and
assign the result back into `names(sqf_agg_wide)`. It might look weird to put
a function inside of a function, but R is completely fine with it. Running
`names(sqf_agg_wide)` at the end will print out the column names so we can
check that it worked.

```
library(janitor)
#
# Attaching package: 'janitor'
# The following objects are masked from 'package:stats':
#
#       chisq.test, fisher.test
names(sqf_agg_wide) <- make_clean_names(names(sqf_agg_wide))
names(sqf_agg_wide)
# [1] "month2"                    "day2"
# [3] "asian_pacific_islander"    "black"
# [5] "black_hispanic"            "white"
# [7] "white_hispanic"            "american_indian_alaskan_n"
```

Now each column name is lowercased and has only underscores instead of
spaces and slashes.

To reshape this wide data to long format, we'll use the `tidyr` function
`pivot_longer()`. There are three inputs here: "cols", which takes a vector of
column names which have our value variables; "names_to", which is what it'll

call the newly created categorical variable; and "values_to", which is what it'll call the newly created values column. For "cols" we want to include each of our race columns, and these column names must be in quotes and in a vector as there are multiple columns. For "names_to" we can call it whatever we want, but here we'll call it "race" as the variable is about the race of the person who was stopped. And for "values_to" we'll call it "number_of_people_stopped" though we can call it whatever we like.

```
sqf_agg_long <- sqf_agg_wide %>%
  pivot_longer(
    cols = c(
      "asian_pacific_islander",
      "black",
      "black_hispanic",
      "white",
      "white_hispanic",
      "american_indian_alaskan_n"
    ),
    names_to = "race",
    values_to = "number_of_people_stopped"
  )
head(sqf_agg_long)
# # A tibble: 6 x 4
# # Groups:   month2, day2 [1]
#   month2 day2   race                       number_of_people~1
#   <chr>  <chr>  <chr>                                   <int>
# 1 April  Friday asian_pacific_islander                      1
# 2 April  Friday black                                     104
# 3 April  Friday black_hispanic                             17
# 4 April  Friday white                                      16
# 5 April  Friday white_hispanic                             31
# 6 April  Friday american_indian_alaskan_n                  NA
# # ... with abbreviated variable name
# #   1: number_of_people_stopped
```

In some cases you'll have many columns that you want to include while reshaping, which makes writing them all out by hand time consuming. If all of the columns are sequential you can use a trick in this function by writing first_column:last_column where the : will make it include each column (in order) from the first one you input to the last one. This is doing the same thing as 1:3, which returns 1, 2, and 3, but for columns instead of numbers. This doesn't work in most cases but does work for many tidyverse packages. There are also a large number of functions from the dplyr package that are for selecting columns, and are very helpful for doing things like this. The functions are numerous and have changed relatively frequently in the past so I won't cover them in this book, but if you're interested you can look at them on this page[3] of `dplyr`'s website.

```
sqf_agg_long <- sqf_agg_wide %>%
  pivot_longer(
    cols = asian_pacific_islander:american_indian_alaskan_n,
    names_to = "race",
    values_to = "number_of_people_stopped"
  )
head(sqf_agg_long)
# # A tibble: 6 x 4
# # Groups:   month2, day2 [1]
#   month2 day2   race                       number_of_people~1
#   <chr>  <chr>  <chr>                                   <int>
# 1 April  Friday asian_pacific_islander                      1
# 2 April  Friday black                                     104
# 3 April  Friday black_hispanic                             17
# 4 April  Friday white                                      16
# 5 April  Friday white_hispanic                             31
# 6 April  Friday american_indian_alaskan_n                  NA
# # ... with abbreviated variable name
# #   1: number_of_people_stopped
```

13.2   Reshaping multiple columns

So far we've just been reshaping using a single column. This is the simplest method, but in some cases we'll need to reshape using multiple columns. As an example, let's make a new column called "n2" in our "sqf_agg" data set, which just adds 10 to the value in our "n" column.

```
sqf_agg$n2 <- sqf_agg$n + 10
```

Since these columns both relate to the race column (called "SUS-PECT_RACE_DESCRIPTION") we can reuse the `pivot_wider()` code from before, but now the `values_from` parameter takes a vector of column names instead of a single column name. Here we want to include both the "n" and the

---

[3]https://dplyr.tidyverse.org/reference/dplyr_tidy_select.html

"n2" column. Since it's a `dplyr` function it's not necessary to put the column names in quotes. We also want to keep our `filter()` function from before which removes "(null)" races and then add a `head()` function at the end so it prints out the first six rows of our resulting data set.

```
sqf_agg_wide <- sqf_agg %>%
  filter(SUSPECT_RACE_DESCRIPTION != "(null)") %>%
  pivot_wider(
    names_from = SUSPECT_RACE_DESCRIPTION,
    values_from = c(n, n2)
  )
names(sqf_agg_wide) <- make_clean_names(names(sqf_agg_wide))
head(sqf_agg_wide)
# # A tibble: 6 x 14
# # Groups:   month2, day2 [6]
#   month2 day2      n_asian_~1 n_black n_bla~2 n_white n_whi~3
#   <chr>  <chr>          <int>   <int>   <int>   <int>   <int>
# 1 April  Friday             1     104      17      16      31
# 2 April  Monday             1      92      10      32      29
# 3 April  Saturday           3     115      24      24      44
# 4 April  Sunday             2      96      12      15      38
# 5 April  Thursday           2     122      10      18      38
# 6 April  Tuesday            7     137      14      20      49
# # ... with 7 more variables:
# #   n_american_indian_alaskan_n <int>,
# #   n2_asian_pacific_islander <dbl>, n2_black <dbl>,
# #   n2_black_hispanic <dbl>, n2_white <dbl>,
# #   n2_white_hispanic <dbl>,
# #   n2_american_indian_alaskan_n <dbl>, and abbreviated
# #   variable names 1: n_asian_pacific_islander, ...
```

We now have the same wide data set as before, but now there are twice as many race columns. And the `pivot_wider()` function renamed the columns so we can tell the "n" columns from the "n2" columns. The easiest way to reshape this data from wide to long is to again use the `pivot_longer()` function but now use it twice: first to reshape the "n" columns and then to reshape the "n2" columns. We'll use the exact same code as before, but change the column names to suit their new names.

```
sqf_agg_long <- sqf_agg_wide %>%
  pivot_longer(
    cols = c(
      "n_asian_pacific_islander",
```

```
      "n_black",
      "n_black_hispanic",
      "n_white",
      "n_white_hispanic",
      "n_american_indian_alaskan_n"
    ),
    names_to = "race",
    values_to = "number_of_people_stopped"
  ) %>%
  pivot_longer(
    cols = c(
      "n2_asian_pacific_islander",
      "n2_black",
      "n2_black_hispanic",
      "n2_white",
      "n2_white_hispanic",
      "n2_american_indian_alaskan_n"
    ),
    names_to = "race2",
    values_to = "number_of_people_stopped2"
  )
head(sqf_agg_long)
# # A tibble: 6 x 6
# # Groups:   month2, day2 [1]
#   month2 day2    race                     numbe~1 race2 numbe~2
#   <chr>  <chr>   <chr>                      <int> <chr>   <dbl>
# 1 April  Friday  n_asian_pacific_islan~         1 n2_a~      11
# 2 April  Friday  n_asian_pacific_islan~         1 n2_b~     114
# 3 April  Friday  n_asian_pacific_islan~         1 n2_b~      27
# 4 April  Friday  n_asian_pacific_islan~         1 n2_w~      26
# 5 April  Friday  n_asian_pacific_islan~         1 n2_w~      41
# 6 April  Friday  n_asian_pacific_islan~         1 n2_a~      NA
# # ... with abbreviated variable names
# #   1: number_of_people_stopped,
# #   2: number_of_people_stopped2
```

This now gives us two race columns - "race" and "race2" - which are ordered differently so we need to make sure to either reorder the data to be the same ordering or to keep that in mind when comparing the "number_of_people_stopped" and "number_of_people_stopped2" columns as they frequently refer to different races.

Part IV

Visualize

14

Graphing with *ggplot2*

For this chapter you'll need the following file, which is available for download here[1]: apparent_per_capita_alcohol_consumption.rda.

We've made some simple graphs earlier; in this lesson we will use the package *ggplot2* to make simple and elegant-looking graphs.

The "gg" part of *ggplot2* stands for "grammar of graphics", which is the idea that most graphs can be made using the same few "pieces." We'll get into those pieces during this lesson. For a useful cheat sheet for this package see here[2].

```
install.packages("ggplot2")
```

```
library(ggplot2)
```

When working with new data, it's often useful to quickly graph the data to try to understand what you're working with. It is also useful when understanding how much to trust the data.

The data we will work on is data about alcohol consumption in US states from 1977-2017 from the National Institutes of Health. It contains the per capita alcohol consumption for each state for every year. Their method to determine per capita consumption is amount of alcohol sold / number of people aged 14+ living in the state. More details on the data are available here[3].

Now we need to load the data.

```
load("data/apparent_per_capita_alcohol_consumption.rda")
```

---

[1] https://github.com/jacobkap/crimebythenumbers/tree/master/data
[2] https://www.rstudio.com/wp-content/uploads/2015/03/ggplot2-cheatsheet.pdf
[3] https://www.openicpsr.org/openicpsr/project/105583/version/V2/view

The name of the data is quite long so for convenience let's copy it to a new object with a better name, *alcohol*.

```
alcohol <- apparent_per_capita_alcohol_consumption
```

The original data has every state, region, and the US as a whole. For this lesson we're using data subsetted to just include states. For now let's just look at Pennsylvania.

```
penn_alcohol <- alcohol[alcohol$state == "pennsylvania", ]
```

14.1   What does the data look like?

Before graphing, it's helpful to see what the data includes. An important thing to check is what variables are available and what the units are for these variables.

```
head(penn_alcohol)
#            state year ethanol_beer_gallons_per_capita
# 1559 pennsylvania 2017                            1.29
# 1560 pennsylvania 2016                            1.31
# 1561 pennsylvania 2015                            1.31
# 1562 pennsylvania 2014                            1.32
# 1563 pennsylvania 2013                            1.34
# 1564 pennsylvania 2012                            1.36
#      ethanol_wine_gallons_per_capita
# 1559                            0.33
# 1560                            0.33
# 1561                            0.32
# 1562                            0.32
# 1563                            0.31
# 1564                            0.31
#      ethanol_spirit_gallons_per_capita
# 1559                              0.71
# 1560                              0.72
# 1561                              0.70
# 1562                              0.70
```

```
# 1563                                  0.68
# 1564                                  0.67
#        ethanol_all_drinks_gallons_per_capita number_of_beers
# 1559                                    2.34        305.7778
# 1560                                    2.36        310.5185
# 1561                                    2.33        310.5185
# 1562                                    2.34        312.8889
# 1563                                    2.33        317.6296
# 1564                                    2.34        322.3704
#        number_of_glasses_wine number_of_shots_liquor
# 1559                 65.48837               147.4128
# 1560                 65.48837               149.4891
# 1561                 63.50388               145.3366
# 1562                 63.50388               145.3366
# 1563                 61.51938               141.1841
# 1564                 61.51938               139.1079
#        number_of_drinks_total
# 1559                 499.2000
# 1560                 503.4667
# 1561                 497.0667
# 1562                 499.2000
# 1563                 497.0667
# 1564                 499.2000
```

So each row of the data is a single year of data for Pennsylvania. It includes alcohol consumption for wine, liquor, beer, and total drinks - both as gallons of ethanol (a hard unit to interpret) and more traditional measures such as glasses of wine or number of beers. The original data only included the gallons of ethanol data, which I converted to the more understandable units. If you encounter data with odd units, it is a good idea to convert it to something easier to understand - especially if you intend to show someone else the data or results.

14.2   Graphing data

To make a plot using `ggplot()` (please note that the function does not have a 2 at the end of it, only the package name does), all you need to do is specify the data set and the variables you want to plot. From there you add on pieces of the graph using the + symbol (which operates like a `dplyr` pipe) and then specify what you want added.

For ggplot() we need to specify four things:

1. The data set
2. The x-axis variable
3. The y-axis variable
4. The type of graph - e.g. line, point, etc.

Some useful types of graphs are:

-   geom_point() - A point graph, can be used for scatter plots
-   geom_line() - A line graph
-   geom_bar() - A barplot
-   geom_smooth() - Adds a regression line to the graph

14.3 Time-series plots

Let's start with a time-series of beer consumption in Pennsylvania. In time-series plots the x-axis is always the time variable while the y-axis is the variable whose trend over time is what we're interested in. When you see a graph showing, for example, crime rates over time, this is the type of graph you're looking at.

The code below starts by writing our data set name. Then says what our x- and y-axis variables are called. The x- and y-axis variables are within parentheses of the function called aes(). aes() stands for aesthetic, and what's included inside here describes how the graph will look. It's not intuitive to remember, but you need to include it. Like in dplyr functions, you do not need to put the column names in quotes or repeat which data set you are using.

```
ggplot(penn_alcohol, aes(
  x = year,
  y = number_of_beers
))
```

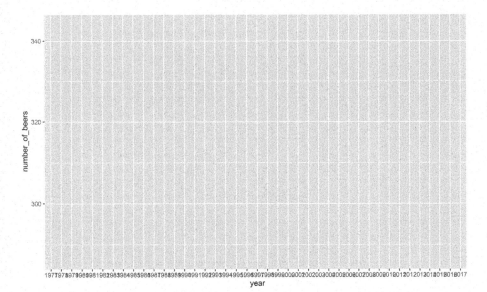

Note that on the x-axis it prints out every single year and makes it completely unreadable. That is because the "year" column is a character type, so R thinks each year is its own category. It prints every single year because it thinks we want every category shown. To fix this, we can make the column numeric, and `ggplot()` will be smarter about printing fewer years.

```
penn_alcohol$year <- as.numeric(penn_alcohol$year)
```

```
ggplot(penn_alcohol, aes(
  x = year,
  y = number_of_beers
))
```

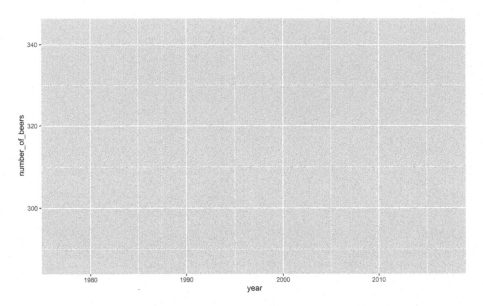

When we run it, we get our graph. It includes the variable names for each axis and shows the range of data through the tick marks. What is missing is the actual data. For that we need to specify what type of graph it is. We literally add it with the + followed by the type of graph we want. Make sure that the + is at the end of a line, not the start of one. Starting a line with the + will not work.

Let's start with point and line graphs.

```
ggplot(penn_alcohol, aes(
  x = year,
  y = number_of_beers
)) +
  geom_point()
```

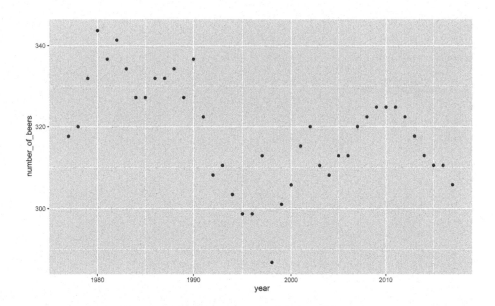

```
ggplot(penn_alcohol, aes(
  x = year,
  y = number_of_beers
)) +
  geom_line()
```

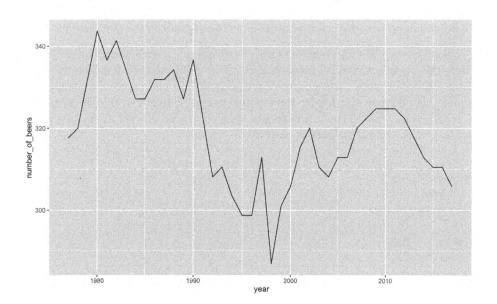

We can also combine different types of graphs.

```
ggplot(penn_alcohol, aes(
  x = year,
  y = number_of_beers
)) +
  geom_point() +
  geom_line()
```

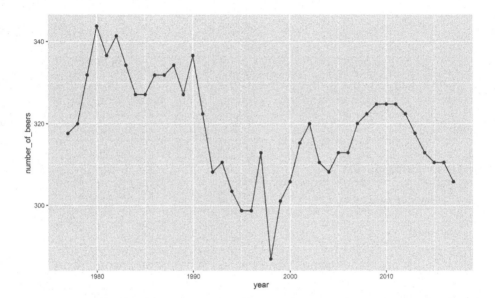

It looks like there's a huge change in beer consumption over time. But look at where they y-axis starts. It starts around 280 so really that change is only ~60 beers. That's because when graphs don't start at 0, it can make small changes appear big. We can fix this by forcing the y-axis to begin at 0. We can add `expand_limits(y = 0)` to the graph to say that the value 0 must always appear on the y-axis, even if no data is close to that value.

```
ggplot(penn_alcohol, aes(
  x = year,
  y = number_of_beers
)) +
  geom_point() +
  geom_line() +
  expand_limits(y = 0)
```

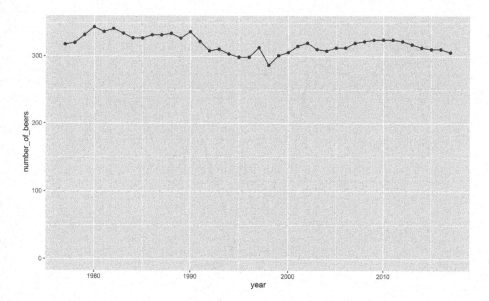

Now that graph shows what looks like nearly no change even though that is also not true. Which graph is best? It's hard to say.

Inside the types of graphs we can change how it is displayed. As with using `plot()`, we can specify the color and size of our lines or points.

```
ggplot(penn_alcohol, aes(
    x = year,
    y = number_of_beers
)) +
    geom_line(color = "forestgreen", size = 1.3)
```

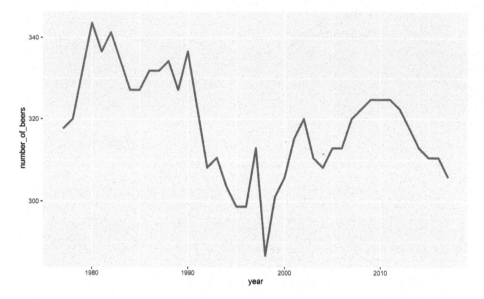

Some other useful features are changing the axis labels and the graph title. Unlike in `plot()` we do not include it in the () of `ggplot()` but use their own functions to add them to the graph. The input to each of these functions is a string for what we want it to say.

- `xlab()` - x-axis label
- `ylab()` - y-axis label
- `ggtitle()` - graph title

```
ggplot(penn_alcohol, aes(
  x = year,
  y = number_of_beers
)) +
  geom_line(color = "forestgreen", size = 1.3) +
  xlab("Year") +
  ylab("Number of Beers") +
  ggtitle("PA Annual Beer Consumption Per Capita (1977-2017)")
```

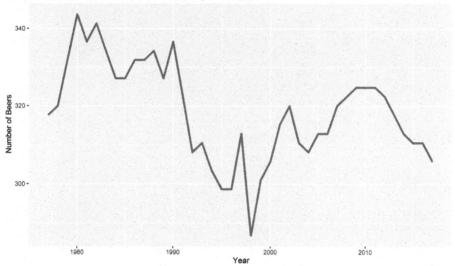

Many time-series plots show multiple variables over the same time period (e.g. murder and robbery over time). There are ways to change the data itself to make creating graphs like this easier, but let's stick with the data we currently have and just change `ggplot()`.

Start with a normal line graph, this time looking at wine.

```
ggplot(penn_alcohol, aes(
  x = year,
  y = number_of_glasses_wine
)) +
  geom_line()
```

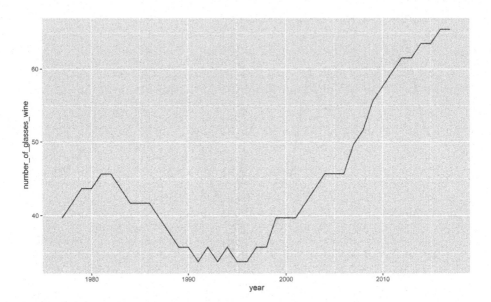

Then include a second `geom_line()` with its own `aes()` for the second variable. Since we are using the penn_alcohol data set for both lines we do not need to include it in the second `geom_line()` as it assumes that the data is the same if we don't specify otherwise. If we used a different data set for the second line, we would need to specify which data set it is inside of `geom_line()` and before `aes()`.

```
ggplot(penn_alcohol, aes(
   x = year,
   y = number_of_glasses_wine
)) +
   geom_line() +
   geom_line(aes(
     x = year,
     y = number_of_shots_liquor
   ))
```

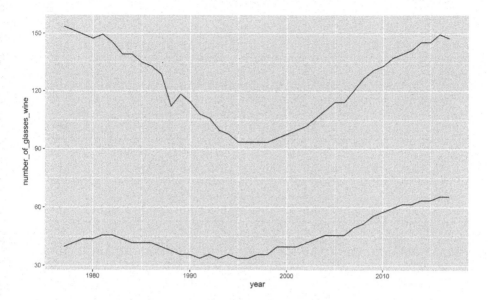

A problem with this is that both lines are the same color. We need to set a color for each line and do so within `aes()`. Instead of providing a color name, we need to provide the name the color will have in the legend. Do so for both lines.

```
ggplot(penn_alcohol, aes(
  x = year,
  y = number_of_glasses_wine,
  color = "Glasses of Wine"
)) +
  geom_line() +
  geom_line(aes(
    x = year,
    y = number_of_shots_liquor,
    color = "Shots of Liquor"
  ))
```

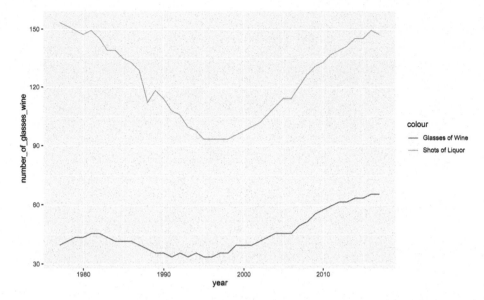

We can change the legend title by using the function labs() and changing the value color to what we want the legend title to be.

```
ggplot(penn_alcohol, aes(
    x = year,
    y = number_of_glasses_wine,
    color = "Glasses of Wine"
)) +
    geom_line() +
    geom_line(aes(
        x = year,
        y = number_of_shots_liquor,
        color = "Shots of Liquor"
    )) +
    labs(color = "Alcohol Type")
```

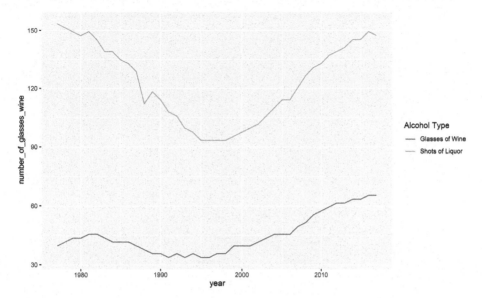

Finally, a useful option to move the legend from the side to the bottom is setting the `theme()` function to move the `legend.position` to "bottom". This will allow the graph to be wider.

```
ggplot(penn_alcohol, aes(
  x = year,
  y = number_of_glasses_wine,
  color = "Glasses of Wine"
)) +
  geom_line() +
  geom_line(aes(
    x = year,
    y = number_of_shots_liquor,
    color = "Shots of Liquor"
  )) +
  labs(color = "Alcohol Type") +
  theme(legend.position = "bottom")
```

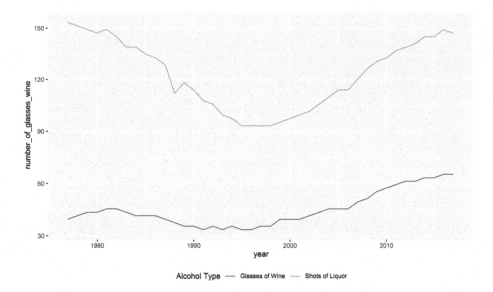

14.4   Scatter plots

Making a scatter plot simply requires changing the x-axis from year to another
numerical variable and using geom_point(). Since our data has one row for every
year for Pennsylvania, we can make a scatterplot comparing different drinks
in each year. For this example, we'll compare liquor to beer sales.

```
ggplot(penn_alcohol, aes(
    x = number_of_shots_liquor,
    y = number_of_beers
)) +
    geom_point()
```

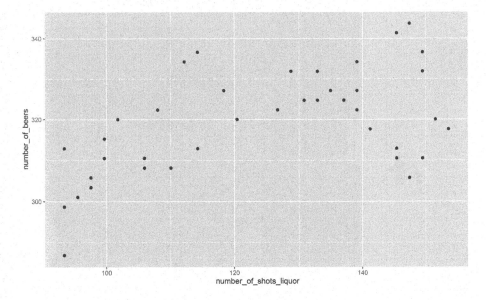

This graph shows us that when liquor consumption increases, beer consumption also tends to increase.

While scatterplots can help show the relationship between variables, we lose the information of how consumption changes over time.

14.5   Color blindness

Please keep in mind that some people are color blind so graphs (or maps, which we will learn about soon) will be hard to read for these people if we choose bad colors. A helpful site for choosing colors for graphs and maps is Color Brewer.[4]

---

[4]http://colorbrewer2.org

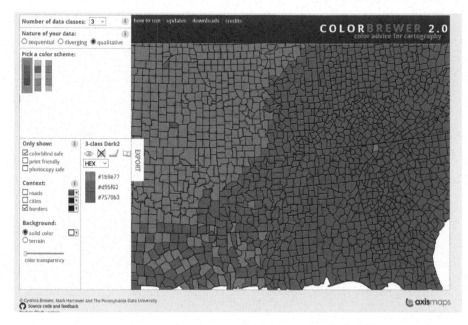

This site lets you select which type of colors you want (sequential and diverging, such as shades in a hotspot map, and qualitative, such as for data like what we used in this lesson). In the "Only show:" section you can set it to "colorblind safe" to restrict it to colors that allow people with color blindness to read your graph. To the right of this section it shows the HEX codes for each color. A HEX code is just a code that a computer can read and know exactly which color it is.

Let's use an example of a color-blind friendly color from the "qualitative" section of ColorBrewer. We have three options on this page (we can change how many colors we want but it defaults to showing 3): green (HEX = #1b9e77), orange (HEX = #d95f02), and purple (HEX = #7570b3). We'll use the orange and purple colors. To manually set colors in `ggplot()` we use `scale_color_manual(values = c())` and include a vector of color names or HEX codes inside the `c()`. Doing that using the orange and purple HEX codes will change our graph colors to these two colors.

```
ggplot(penn_alcohol, aes(
  x = year,
  y = number_of_glasses_wine,
  color = "Glasses of Wine"
)) +
  geom_line() +
  geom_line(aes(
    x = year,
```

```
      y = number_of_shots_liquor,
      color = "Shots of Liquor"
)) +
labs(color = "Alcohol Type") +
theme(legend.position = "bottom") +
scale_color_manual(values = c("#7570b3", "#d95f02"))
```

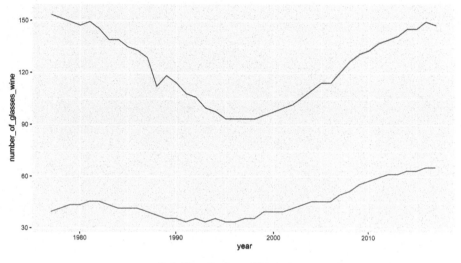

15

More graphing with `ggplot2`

For this chapter you'll need the following file, which is available for download here[1]: fatal-police-shootings-data.csv.

In this lesson we will continue to explore graphing using `ggplot2`. The data we will use is microdata on officer-involved shootings that resulted in a death in the United States since January 1st, 2015. This data has been compiled and released by *The Washington Post* so it will be a useful exercise in exploring data from non-government sources. This data is useful for our purposes as it has a number of variables related to the person who was shot, allowing us to practice making many types of graphs. Each row of data is a different person who was shot and killed by the police, and each column gives us information about the individual or the shooting, such as their age, whether they carried any weapon, and the shooting location.

To explore the data on *The Washington Post's* website, see here.[2] To examine their methodology, see here.[3]

The data initially comes as a .csv file so we'll use the `read_csv()` function from the `readr` package.

```
library(readr)
shootings <- read_csv("data/fatal-police-shootings-data.csv")
```

Since `read_csv()` reads files into a tibble object, we'll turn it into a data.frame so `head()` shows every single column.

```
shootings <- as.data.frame(shootings)
```

---

[1]https://github.com/jacobkap/crimebythenumbers/tree/master/data

[2]https://www.washingtonpost.com/graphics/2019/national/police-shootings-2019/?ut m_term=.e870afc9a00c

[3]https://www.washingtonpost.com/national/how-the-washington-post-is-examining-police-shootings-in-the-united-states/2016/07/07/d9c52238-43ad-11e6-8856-f26de2537a9d _story.html?utm_term=.f07e9800092b

15.1   Exploring data

Now that we have the data read in, let's look at it.

```
ncol(shootings)
# [1] 14
nrow(shootings)
# [1] 4371
```

The data has 14 variables and covers 4,371 shootings. Let's check out some of the variables, first using head() then using summary() and table().

```
head(shootings)
#   id               name       date  manner_of_death
# 1  3        Tim Elliot 2015-01-02             shot
# 2  4  Lewis Lee Lembke 2015-01-02             shot
# 3  5 John Paul Quintero 2015-01-03 shot and Tasered
# 4  8   Matthew Hoffman 2015-01-04             shot
# 5  9 Michael Rodriguez 2015-01-04             shot
# 6 11 Kenneth Joe Brown 2015-01-04             shot
#        armed age gender race          city state
# 1        gun  53      M    A       Shelton    WA
# 2        gun  47      M    W        Aloha    OR
# 3    unarmed  23      M    H      Wichita    KS
# 4 toy weapon  32      M    W San Francisco    CA
# 5   nail gun  39      M    H        Evans    CO
# 6        gun  18      M    W      Guthrie    OK
#   signs_of_mental_illness threat_level        flee
# 1                    TRUE       attack Not fleeing
# 2                   FALSE       attack Not fleeing
# 3                   FALSE        other Not fleeing
# 4                    TRUE       attack Not fleeing
# 5                   FALSE       attack Not fleeing
# 6                   FALSE       attack Not fleeing
#   body_camera
# 1       FALSE
# 2       FALSE
# 3       FALSE
# 4       FALSE
```

```
# 5          FALSE
# 6          FALSE
```

Each row is a single shooting, and it includes variables such as the victim's name, the date of the shooting, demographic information about that person, the city and state where the shooting occurred, and some information about the incident. It is clear from these first 6 rows that most variables are categorical so we can't use summary() on them. Let's use summary() on the date and age columns and then use table() for the rest.

```
summary(shootings$date)
#         Min.      1st Qu.       Median         Mean
# "2015-01-02" "2016-02-07" "2017-03-16" "2017-03-18"
#      3rd Qu.         Max.
# "2018-04-11" "2019-06-25"
summary(shootings$age)
#   Min. 1st Qu.  Median    Mean 3rd Qu.    Max.    NA's
#   6.00   27.00   35.00   36.84   45.00   91.00     182
```

From this we can see that the data is from early January through mid-2019.[4] From the age column we can see that the average age is about 37 with most people around that range. Now we can use table() to see how often each value appears in each variable. We don't want to do this for city or name as there would be too many values, but it will work for the other columns. Let's start with the "manner_of_death" column.

```
table(shootings$manner_of_death)
#
#            shot shot and Tasered
#            4146              225
```

To turn these counts into percentages we can divide the results by the number of rows in our data and multiply by 100.

```
table(shootings$manner_of_death) / nrow(shootings) * 100
#
#              shot shot and Tasered
#         94.852437         5.147563
```

---

[4] *The Washington Post* is continuing to collect this data so if you look on their site you'll find more up-to-date data.

Now it is clear to see that in about 95% of shootings, officers used a gun and in 5% of shootings they also used a Taser. As this is data on officer shooting deaths, this is unsurprising. Let's take a look at whether the victim was armed.

```
table(shootings$armed) / nrow(shootings) * 100
#
#                           air conditioner
#                                0.02287806
#                                        ax
#                                0.48043926
#                                  barstool
#                                0.02287806
#                              baseball bat
#                                0.27453672
#                  baseball bat and bottle
#                                0.02287806
# baseball bat and fireplace poker
#                                0.02287806
#                                     baton
#                                0.09151224
#                                   bayonet
#                                0.02287806
#                                    BB gun
#                                0.06863418
#                              bean-bag gun
#                                0.02287806
#                               beer bottle
#                                0.06863418
#                              blunt object
#                                0.11439030
#                             bow and arrow
#                                0.02287806
#                                box cutter
#                                0.22878060
#                                     brick
#                                0.04575612
#                                   carjack
#                                0.02287806
#                                     chain
#                                0.04575612
#                                  chain saw
#                                0.04575612
#                                  chainsaw
#                                0.02287806
```

```
#                         chair
#                    0.04575612
#            claimed to be armed
#                    0.02287806
#            contractor's level
#                    0.02287806
#                 cordless drill
#                    0.02287806
#                       crossbow
#                    0.20590254
#                        crowbar
#                    0.06863418
#                      fireworks
#                    0.02287806
#                       flagpole
#                    0.02287806
#                     flashlight
#                    0.02287806
#                    garden tool
#                    0.02287806
#                    glass shard
#                    0.06863418
#                            gun
#                   55.43353924
#                    gun and car
#                    0.11439030
#                  gun and knife
#                    0.34317090
#                  gun and sword
#                    0.02287806
#                gun and vehicle
#                    0.04575612
#           guns and explosives
#                    0.06863418
#                         hammer
#                    0.22878060
#                     hand torch
#                    0.02287806
#                        hatchet
#                    0.18302448
#                hatchet and gun
#                    0.04575612
#              incendiary device
#                    0.04575612
```

```
#                          knife
#                    14.96225120
#              lawn mower blade
#                     0.04575612
#                        machete
#                     0.86936628
#               machete and gun
#                     0.02287806
#                   meat cleaver
#                     0.06863418
#               metal hand tool
#                     0.02287806
#                   metal object
#                     0.09151224
#                     metal pipe
#                     0.25165866
#                     metal pole
#                     0.04575612
#                     metal rake
#                     0.02287806
#                    metal stick
#                     0.06863418
#                     motorcycle
#                     0.02287806
#                       nail gun
#                     0.02287806
#                            oar
#                     0.02287806
#                     pellet gun
#                     0.02287806
#                            pen
#                     0.02287806
#                   pepper spray
#                     0.02287806
#                       pick-axe
#                     0.06863418
#                  piece of wood
#                     0.06863418
#                           pipe
#                     0.13726836
#                      pitchfork
#                     0.04575612
#                           pole
#                     0.04575612
```

```
#                   pole and knife
#                     0.04575612
#                            rock
#                     0.09151224
#                   samurai sword
#                     0.02287806
#                        scissors
#                     0.06863418
#                     screwdriver
#                     0.18302448
#                     sharp object
#                     0.11439030
#                          shovel
#                     0.06863418
#                           spear
#                     0.02287806
#                         stapler
#                     0.02287806
#              straight edge razor
#                     0.06863418
#                           sword
#                     0.34317090
#                           Taser
#                     0.41180508
#                       tire iron
#                     0.02287806
#                      toy weapon
#                     3.54609929
#                         unarmed
#                     6.36010066
#                    undetermined
#                     4.30107527
#                  unknown weapon
#                     1.25829330
#                         vehicle
#                     1.57858614
#                 vehicle and gun
#                     0.02287806
#                    walking stick
#                     0.02287806
#                          wrench
#                     0.02287806
```

This is fairly hard to interpret as it is sorted alphabetically when we'd prefer it to be sorted by most common weapon. It also doesn't round the numbers so there are many numbers past the decimal point shown. Let's solve these two issues using sort() and round(). We could just wrap our initial code inside each of these functions but to avoid making too complicated code, we save the results in a vector called "temp" and incrementally use sort() and round() on that. We'll set the parameter decreasing to TRUE in the sort() function so that it is in descending order of how common each value is. And we'll round to two decimal places by setting the parameter digits to 2 in round().

```
temp <- table(shootings$armed) / nrow(shootings) * 100
temp <- sort(temp, decreasing = TRUE)
temp <- round(temp, digits = 2)
temp
#
#                            gun
#                          55.43
#                          knife
#                          14.96
#                        unarmed
#                           6.36
#                   undetermined
#                           4.30
#                     toy weapon
#                           3.55
#                        vehicle
#                           1.58
#                unknown weapon
#                           1.26
#                        machete
#                           0.87
#                             ax
#                           0.48
#                          Taser
#                           0.41
#                  gun and knife
#                           0.34
#                          sword
#                           0.34
#                   baseball bat
#                           0.27
#                     metal pipe
#                           0.25
#                     box cutter
```

```
#                    0.23
#                   hammer
#                    0.23
#                  crossbow
#                    0.21
#                  hatchet
#                    0.18
#                screwdriver
#                    0.18
#                    pipe
#                    0.14
#                blunt object
#                    0.11
#                gun and car
#                    0.11
#                sharp object
#                    0.11
#                    baton
#                    0.09
#                metal object
#                    0.09
#                    rock
#                    0.09
#                   BB gun
#                    0.07
#                beer bottle
#                    0.07
#                  crowbar
#                    0.07
#                glass shard
#                    0.07
#              guns and explosives
#                    0.07
#                meat cleaver
#                    0.07
#                metal stick
#                    0.07
#                  pick-axe
#                    0.07
#                piece of wood
#                    0.07
#                  scissors
#                    0.07
#                   shovel
```

```
#                                0.07
#                    straight edge razor
#                                0.07
#                               brick
#                                0.05
#                               chain
#                                0.05
#                            chain saw
#                                0.05
#                               chair
#                                0.05
#                       gun and vehicle
#                                0.05
#                      hatchet and gun
#                                0.05
#                     incendiary device
#                                0.05
#                     lawn mower blade
#                                0.05
#                           metal pole
#                                0.05
#                            pitchfork
#                                0.05
#                                pole
#                                0.05
#                        pole and knife
#                                0.05
#                        air conditioner
#                                0.02
#                             barstool
#                                0.02
#                   baseball bat and bottle
#                                0.02
# baseball bat and fireplace poker
#                                0.02
#                              bayonet
#                                0.02
#                          bean-bag gun
#                                0.02
#                          bow and arrow
#                                0.02
#                              carjack
#                                0.02
#                             chainsaw
```

```
#                         0.02
#          claimed to be armed
#                         0.02
#           contractor's level
#                         0.02
#               cordless drill
#                         0.02
#                    fireworks
#                         0.02
#                     flagpole
#                         0.02
#                   flashlight
#                         0.02
#                  garden tool
#                         0.02
#                gun and sword
#                         0.02
#                   hand torch
#                         0.02
#               machete and gun
#                         0.02
#              metal hand tool
#                         0.02
#                   metal rake
#                         0.02
#                   motorcycle
#                         0.02
#                     nail gun
#                         0.02
#                          oar
#                         0.02
#                   pellet gun
#                         0.02
#                          pen
#                         0.02
#                 pepper spray
#                         0.02
#                samurai sword
#                         0.02
#                        spear
#                         0.02
#                      stapler
#                         0.02
#                    tire iron
```

```
#                               0.02
#                    vehicle and gun
#                               0.02
#                       walking stick
#                               0.02
#                             wrench
#                               0.02
```

Now it is a little easier to interpret. In over half of the cases the victim was carrying a gun. A knife was involved 15% of the time. And 6% of the time they were unarmed. In 4% of cases there is no data on any weapon. That leaves about 20% of cases where one of the many rare weapons were used, including some that overlap with one of the more common categories.

Think about how you'd graph this data. There are 85 unique values in this column though fewer than ten of them are common enough to appear more than 1% of the time. Should we graph all of them? No, that would overwhelm any graph. For a useful graph we would need to combine many of these into a single category - possibly called "other weapons." And how do we deal with values where they could meet multiple larger categories? There is not always a clear answer for these types of questions. It depends on what data you're interested in, the goal of the graph, the target audience, and personal preference.

Let's keep exploring the data by looking at gender and race.

```
table(shootings$gender) / nrow(shootings) * 100
#
#          F          M
#   4.667124 95.218485
```

Nearly all of the shootings are of a man. Given that we saw most shootings involved a person with a weapon and that most violent crimes are committed by men, this shouldn't be too surprising.

```
temp <- table(shootings$race) / nrow(shootings) * 100
temp <- sort(temp)
temp <- round(temp, digits = 2)
temp
#
#      O      N      A      H      B      W
#   0.87   1.46   1.62  16.45  22.90  44.89
```

White people are the largest race group that is killed by police, followed by Black people and Hispanic people. In fact, there are about twice as many White people killed than Black people killed, and about 2.5 times as many White people killed than Hispanic people killed. Does this mean that the oft-repeated claim that Black people are killed at disproportionate rates is wrong? No. This data simply shows the number of people killed; it doesn't give any indication on rates of death per group. You'd need to merge it with Census data to get population to determine a rate per race group. And even that would be insufficient since people are, for example, stopped by police at different rates.

This data provides a lot of information on people killed by the police, but even so it is insufficient to answer many of the questions on that topic. It's important to understand the data not only to be able to answer questions about it, but to know what questions you can't answer - and you'll find when using criminology data that there are a *lot* of questions that you can't answer.[5]

One annoying thing with the gender and race variables is that they don't spell out the name. Instead of "Female", for example, it has "F". For our graphs we want to spell out the words so it is clear to viewers. We'll fix this issue, and the issue of having many weapon categories, as we graph each variable.

15.2 Graphing a single numeric variable

We've spent some time looking at the data so now we're ready to make the graphs. We need to load the ggplot2 package if we haven't done so already this session (i.e. since you last closed RStudio).

```
library(ggplot2)
```

As a reminder, the benefit of using ggplot() is that we can start with a simple plot and build our way up to more complicated graphs. We'll start here by building some graphs to depict a numeric variable - in this case the "age" column. We start every ggplot() the same, by inserting the data set first and then put our x and y variables inside of the aes() parameter. In this case we're

---

[5]It is especially important to not overreach when trying to answer a question when the data can't do it well. Often, no answer is better than a wrong one - especially in a field with serious consequences like criminology. This isn't to say that you should never try to answer questions since no data is perfect and you may be wrong. You should try to develop a deep understanding of the data and be confident that you can actually answer those questions with confidence.

only going to be plotting an x variable so we don't need to write anything for y.

```
ggplot(shootings, aes(x = age))
```

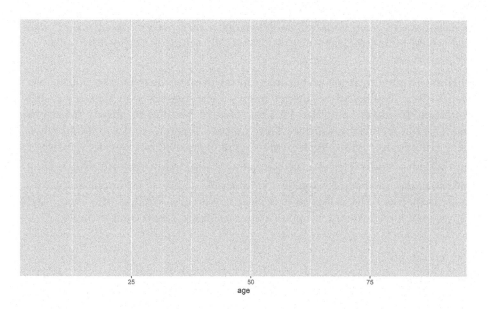

Running the above code returns a blank graph since we haven't told `ggplot()` what type of graph we want yet. Below are a few different types of ways to display a single numeric variable. They're essentially all variations of each other and show the data at different levels of precision. It's hard to say which is best - you'll need to use your best judgment and consider your audience.

15.2.1 Histogram

The histogram is a very common type of graph for a single numeric variable. Histograms group a numeric variable into categories and then plot them, with the heights of each bar indicating how common the group is. We can make a histogram by adding `geom_histogram()` to the `ggplot()`. The results will print out the text "Warning: Removed 182 rows containing non-finite values (stat_bin)." which just means that 182 rows in our data have NA for age, and are excluded from the graph.

```
ggplot(shootings, aes(x = age)) +
  geom_histogram()
# `stat_bin()` using `bins = 30`. Pick better value with
# `binwidth`.
# Warning: Removed 182 rows containing non-finite values
# (stat_bin).
```

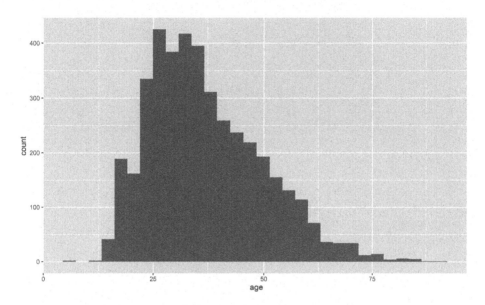

The x-axis is ages with each bar being a group of certain ages, and the y-axis is how many people are in each group. The grouping is done automatically and we can alter it by changing the `bin` parameter in `geom_histogram()`. By default this parameter is set to 30, but we can make each group smaller (have fewer ages per group) by **increasing** it from 30 or make each group larger by **decreasing** it.

```
ggplot(shootings, aes(x = age)) +
  geom_histogram(bins = 15)
# Warning: Removed 182 rows containing non-finite values
# (stat_bin).
```

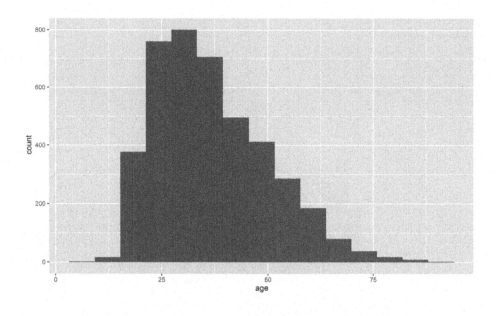

```
ggplot(shootings, aes(x = age)) +
  geom_histogram(bins = 45)
# Warning: Removed 182 rows containing non-finite values
# (stat_bin).
```

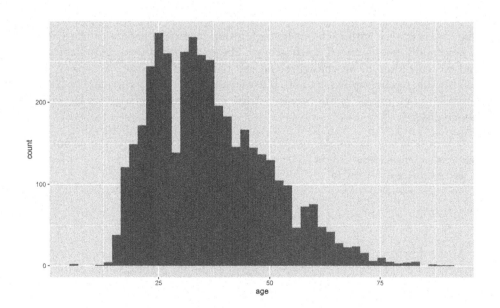

Note that while the overall trend (of most deaths being around age 25) doesn't change when we alter bin, the data gets more or less precise. Having fewer bins means fewer, but larger, bars which can obscure trends that more, smaller, bars would show. But having too many bars may make you focus on minor variations that could occur randomly and take away attention from the overall trend. I prefer to err on the side of more precise graphs (more, smaller bars) but be careful over-interpreting data from small groups.

These graphs show the y-axis as the number of people in each bar. If we want to show proportions instead, we can add in a parameter for y in the aes() of the geom_histogram(). We add in y = (..count..)/sum(..count..)), which automatically converts the counts to proportions. The (..count..)/sum(..count..)) stuff is just taking each group and dividing it from the sum of all groups. You could, of course, do this yourself before making the graph, but it's an easy helper. If you do this, make sure to relabel the y-axis so you don't accidentally call the proportions a count.

```
ggplot(shootings, aes(x = age)) +
  geom_histogram(aes(y = (..count..) / sum(..count..)))
# `stat_bin()` using `bins = 30`. Pick better value with
# `binwidth`.
# Warning: Removed 182 rows containing non-finite values
# (stat_bin).
```

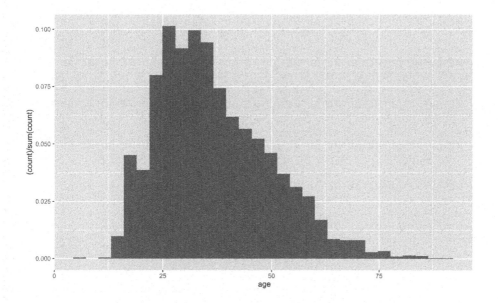

15.2.2   Density plot

Density plots are essentially smoothed versions of histograms. They're especially useful for numeric variables, which are not integers (integers are whole numbers). They're also useful when you want to be more precise than a histogram as they are - to simplify - histograms where each bar is very narrow. Note that the y-axis of a density plot is automatically labeled "density" and has very small numbers. Interpreting the y-axis is fairly hard to explain to someone not familiar with statistics so I'd caution against using this graph unless your audience is already familiar with it.

To interpret these kinds of graphs, I recommend looking for trends rather than trying to identify specific points. For example, in the below graph we can see that shootings rise rapidly starting around age 10, peak at around age 30 (if we were presenting this graph to other people we'd probably want more ages shown on the x-axis), and then steadily decline until about age 80 where it's nearly flat.

```
ggplot(shootings, aes(x = age)) +
  geom_density()
# Warning: Removed 182 rows containing non-finite values
# (stat_density).
```

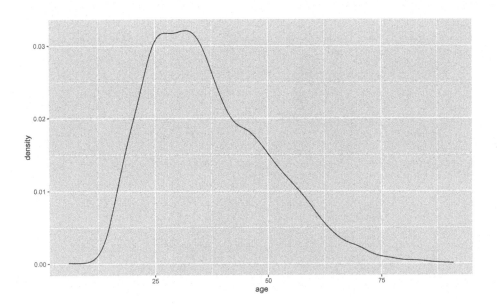

15.2.3 Count graph

A count graph is essentially a histogram with a bar for every value in the numeric variable - like a less-smooth density plot. Note that this won't work well if you have too many unique values so I'd strongly recommend rounding the data to the nearest whole number first if you don't already have an integer. Our age variable is already rounded so we don't need to do that. To make a count graph, we add `stat_count()` to the `ggplot()`.

```
ggplot(shootings, aes(x = age)) +
  stat_count()
# Warning: Removed 182 rows containing non-finite values
# (stat_count).
```

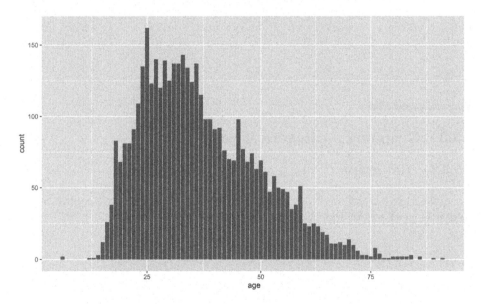

Now we have a single bar for every age in the data. Like the histogram, the y-axis shows the number of people that are that age. And like the histogram, we can change this from number of people to proportion of people using the exact same code.

```
ggplot(shootings, aes(x = age)) +
  stat_count(aes(y = (..count..) / sum(..count..)))
# Warning: Removed 182 rows containing non-finite values
# (stat_count).
```

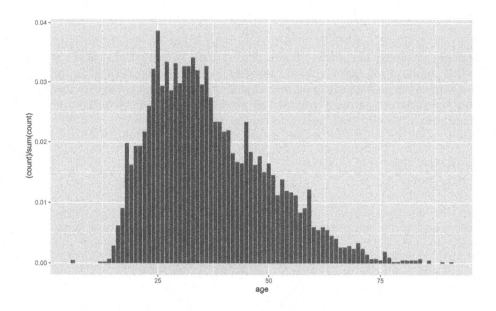

15.3   Graphing a categorical variable

15.3.1   Bar graph

To make this barplot we'll set the x-axis variable to our "race" column and
add geom_bar() to the end.

```
ggplot(shootings, aes(x = race)) +
  geom_bar()
```

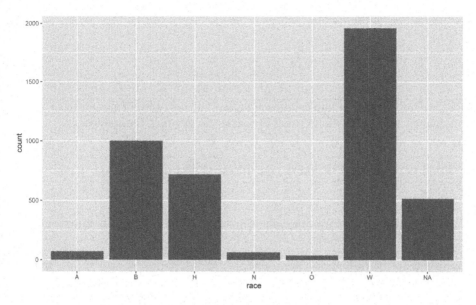

This gives us a barplot in alphabetical order. In most cases we want the data sorted by frequency, so we can easily see which value is the most common, second most common, etc. There are a few ways to do this, but we'll do this by turning the "race" variable into a factor and ordering it by frequency. We can do that using the factor() function. The first input will be the "race" variable, and then we will need to set the levels parameter to a vector of values sorted by frequency. An easy way to know how often values are in a column is to use the table() function on that column, such as below.

```
table(shootings$race)
#
#    A    B    H    N    O    W
#   71 1001  719   64   38 1962
```

It's still alphabetical so let's wrap that in a sort() function.

```
sort(table(shootings$race))
#
#    O    N    A    H    B    W
#   38   64   71  719 1001 1962
```

It's sorted from smallest to largest. We usually want to graph from largest to smallest so let's set the parameter decreasing in sort() to TRUE.

```
sort(table(shootings$race), decreasing = TRUE)
#
#    W    B    H    A    N    O
# 1962 1001  719   71   64   38
```

Now, we only need the names of each value, not how often they occur. So we can against wrap this whole thing in `names()` to get just the names.

```
names(sort(table(shootings$race), decreasing = TRUE))
# [1] "W" "B" "H" "A" "N" "O"
```

If we tie it all together, we can make the "race" column into a factor variable.

```
shootings$race <- factor(shootings$race,
  levels = names(sort(table(shootings$race),
    decreasing = TRUE
  ))
)
```

Now let's try that barplot again.

```
ggplot(shootings, aes(x = race)) +
  geom_bar()
```

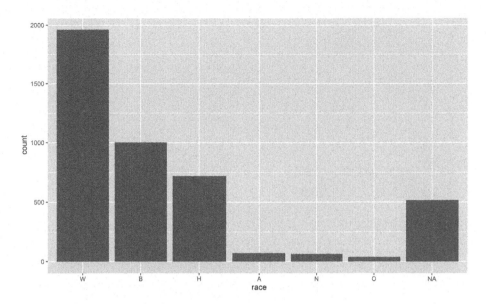

It works! Note that all the values that are missing in our data are still reported in the barplot under a column called "NA." This is not sorted properly since there are more NA values than three of the other values, but NA is still at the far right of the graph. We can change this if we want to make all the NA values an actual character type and call it something like "Unknown." But this way it does draw attention to how many values are missing from this column. Like most things in graphing, this is a personal choice as to what to do.

For bar graphs it is often useful to flip the graph so each value is a row in the graph rather than a column. This also makes it much easier to read the value name. If the value names are long, it'll shrink the graph to accommodate the name. This is usually a sign that you should try to shorten the name to avoid reducing the size of the graph.

```
ggplot(shootings, aes(x = race)) +
  geom_bar() +
  coord_flip()
```

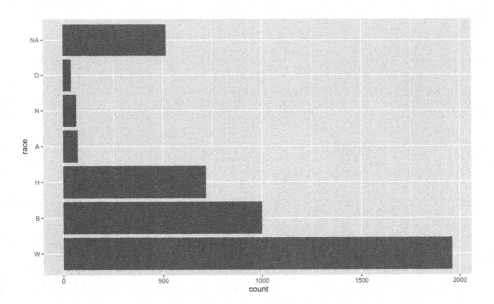

Since it's flipped, now it's sorted from smallest to largest. So we'll need to change the `factor()` code to fix that by making the `decreasing` parameter in `sort()` FALSE.

```
shootings$race <- factor(shootings$race,
  levels = names(sort(table(shootings$race),
    decreasing = FALSE
  ))
)
ggplot(shootings, aes(x = race)) +
  geom_bar() +
  coord_flip()
```

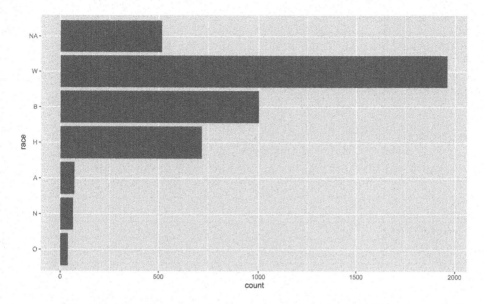

The NA value is now at the top, which looks fairly bad. Let's change all NA values to the string "Unknown". And while we're at it, let's change all the abbreviated race values to actual names. We can get all the NA values by using is.na(shootings$race) and using a conditional statement to get all rows that meet that condition, then assign them the value "Unknown". Instead of trying to subset a factor variable to change the values, we should convert it back to a character type first using as.character(), and then convert it to a factor again once we're done.

```
shootings$race <- as.character(shootings$race)
shootings$race[is.na(shootings$race)] <- "Unknown"
```

Now we can use conditional statements to change all the race letters to names. It's not clear what race "O" and "N" are so I checked *The Washington Post's* GitHub[6] which explains what they mean. Instead of is.na() we'll use shootings$race == "", where we put the letter inside of the quotes.

```
shootings$race[shootings$race == "O"] <- "Other"
shootings$race[shootings$race == "N"] <- "Native American"
shootings$race[shootings$race == "A"] <- "Asian"
shootings$race[shootings$race == "H"] <- "Hispanic"
```

---

[6] https://github.com/washingtonpost/data-police-shootings

```
shootings$race[shootings$race == "B"] <- "Black"
shootings$race[shootings$race == "W"] <- "White"
```

Now let's see how our graph looks. We'll need to rerun the factor() code since now all of the values are changed.

```
shootings$race <- factor(shootings$race,
  levels = names(sort(table(shootings$race),
    decreasing = FALSE
  ))
)
ggplot(shootings, aes(x = race)) +
  geom_bar() +
  coord_flip()
```

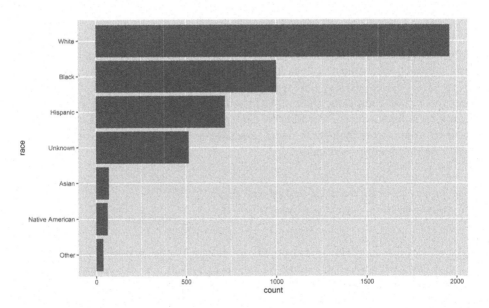

As earlier, we can show proportion instead of count by adding y = (..count..)/sum(..count..) to the aes() in geom_bar().

```
ggplot(shootings, aes(x = race)) +
  geom_bar(aes(y = (..count..) / sum(..count..))) +
  coord_flip()
```

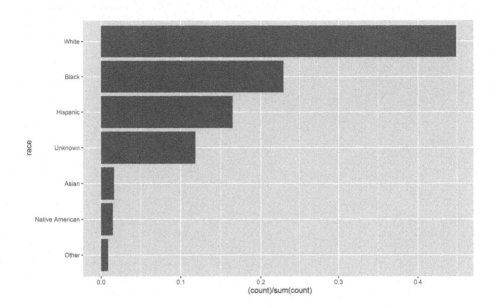

15.4 Graphing data over time

We went over time-series graphs in Chapter 14, but it's such an important topic we'll cover it again. A lot of criminology research is seeing if a policy had an effect, which means we generally want to compare an outcome over time (and compare the treated group to a similar untreated group). To graph that we look at an outcome - in this case numbers of killings - over time. In our case we aren't evaluating any policy, just seeing if the number of police killings change over time.

We'll need to make a variable to indicate that the row is for one shooting. We can call this "dummy" and assign it a value of 1. Then we can make the ggplot() and set this "dummy" column to the y-axis value and set our date variable "date" to the x-axis (the time variable is **always** on the x-axis). Then we'll set the type of plot to geom_line() so we have a line graph showing killings over time.

```
shootings$dummy <- 1
ggplot(shootings, aes(
  x = date,
  y = dummy
```

```
)) +
  geom_line()
```

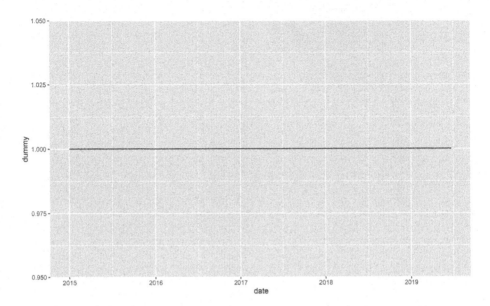

This graph is clearly wrong. Why? Well, our y-axis variable is always 1 so there's no variation to plot. Every single value, even if there are more than one shooting per day, is on the 1 line on the y-axis. And the fact that we have multiple killings per day is an issue because we only want a single line in our graph. We'll need to aggregate our data to some time period (e.g. day, month, year) so that we have one row per time-period and know how many people were killed in that period. We'll start with yearly data and then move to monthly data. Since we're going to be dealing with dates, let's use the lubridate() package that is well-suited for this task.

```
install.packages("lubridate")
```

```
library(lubridate)
#
# Attaching package: 'lubridate'
# The following object is masked from 'package:cowplot':
#
#      stamp
# The following objects are masked from 'package:base':
```

```
#
#        date, intersect, setdiff, union
```

We'll use two functions to create variables that tell us the month and the year of each date in our data. We'll use these new variables to aggregate our data to that time unit. First, the `floor_date()` function is a very useful tool that essentially rounds a date. Here we have the exact date the killing happened on, and we want to determine what month that date is from. So we'll use the parameter `unit` in `floor_date()` and tell the function we want to know the "month" (for a full set of options please see the documentation for `floor_date()` by entering `?floor_date` in the console). So we can do `floor_date(shootings$date, unit = "month")` to get the month - specifically, it returns the date that is the first of the month for that month - in which the killing happened. Even simpler, to get the year, we simply use `year()` and put our "date" variable in the parentheses. We'll call the new variables "month_year" and "year", respectively.

```r
shootings$month_year <- floor_date(shootings$date, unit = "month")
shootings$year <- year(shootings$date)

head(shootings$month_year)
# [1] "2015-01-01" "2015-01-01" "2015-01-01" "2015-01-01"
# [5] "2015-01-01" "2015-01-01"
head(shootings$year)
# [1] 2015 2015 2015 2015 2015 2015
```

Since the data is already sorted by date, all the values printed from `head()` are the same. But you can look at the data using `View()` to confirm that the code worked properly.

We can now aggregate the data by the "month_year" variable and assign the result to a new data set we'll call *monthly_shootings*. We'll use the `group_by()` and `summarize()` functions from `dplyr` that were introduced in Chapter 11 to do this. And we'll use the pipe method of writing `dplyr` code that was discussed in Section 11.4. Since the "dummy" column has a value of 1 for each shooting, we'll sum up this column to get the number of shootings each month/year.

```r
library(dplyr)
monthly_shootings <- shootings %>%
  group_by(month_year) %>%
  summarize(dummy = sum(dummy))
head(monthly_shootings)
```

```
# # A tibble: 6 x 2
#   month_year dummy
#   <date>     <dbl>
# 1 2015-01-01    76
# 2 2015-02-01    77
# 3 2015-03-01    92
# 4 2015-04-01    84
# 5 2015-05-01    71
# 6 2015-06-01    65
```

Since we now have a variable that shows the number of people killed each month, we can graph this new data set. We'll use the same process as earlier, but our data set is now `monthly_shootings` instead of `shootings` and the x-axis variable is "month_year" instead of "date".

```
ggplot(monthly_shootings, aes(
    x = month_year,
    y = dummy
)) +
    geom_line()
```

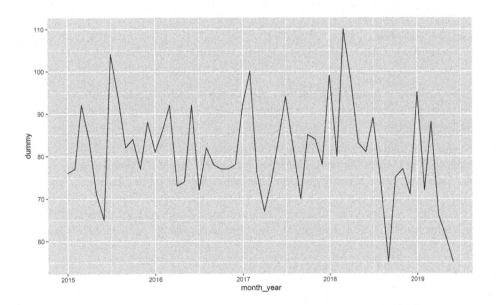

The process is the same for yearly data.

```
yearly_shootings <- shootings %>%
  group_by(year) %>%
  summarize(dummy = sum(dummy))
ggplot(yearly_shootings, aes(
  x = year,
  y = dummy
)) +
  geom_line()
```

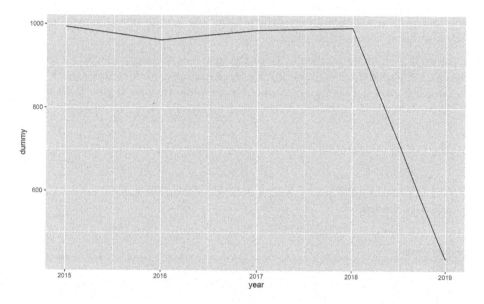

Note the steep drop-off at the end of each graph. Is that due to fewer shooting occurring more recently? No, it's simply an artifact of the graph comparing whole months (years) to parts of a month (year) since we haven't finished this month (year) yet.

15.5 Pretty graphs

What's next for these graphs? You'll likely want to add labels for the axes and the title. We went over how to do this in Section 14.3 so please refer to that for more info. Also, check out `ggplot2`'s website[7] to see more on this very versatile package. As I've said all chapter, a lot of this is going to be personal taste so please spend some time exploring the package and changing the appearance of the graph to learn what looks right to you.

15.5.1   Themes

In addition to making changes to the graph's appearance yourself, you can use a theme that someone else made. A theme is just a collection of changes to the graph's appearance that someone put in a function for others to use. Each theme is different and is fairly opinionated, so you should only use one that you think looks best for your graph. To use a theme, simply add the function for that theme to your ggplot using the + as normal. `ggplot2` comes with a series of themes that you can look at here[8].

Here, we'll be looking at themes from the `ggthemes` package, which is a great source of different themes to modify the appearance of your graph. Check out this website[9] to see a depiction of all of the possible themes. If you don't have the `ggthemes` package installed, do so using `install.packages("ggthemes")`.

```
install.packages("ggthemes")
```

Let's do a few examples using the graph made above. First, we'll need to load the `ggthemes` library.

```
library(ggthemes)
#
# Attaching package: 'ggthemes'
# The following object is masked from 'package:cowplot':
#
#       theme_map
ggplot(yearly_shootings, aes(
  x = year,
  y = dummy
)) +
  geom_line() +
  theme_fivethirtyeight()
```

---

[7] https://ggplot2.tidyverse.org/reference/index.html#section-scales
[8] https://ggplot2.tidyverse.org/reference/ggtheme.html
[9] https://yutannihilation.github.io/allYourFigureAreBelongToUs/ggthemes/

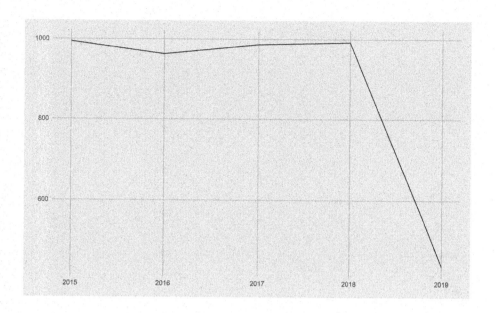

```
ggplot(yearly_shootings, aes(
  x = year,
  y = dummy
)) +
  geom_line() +
  theme_tufte()
```

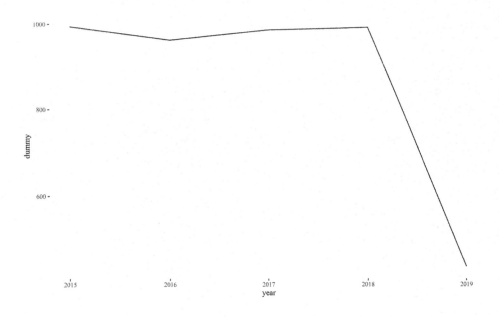

```
ggplot(yearly_shootings, aes(
  x = year,
  y = dummy
)) +
  geom_line() +
  theme_few()
```

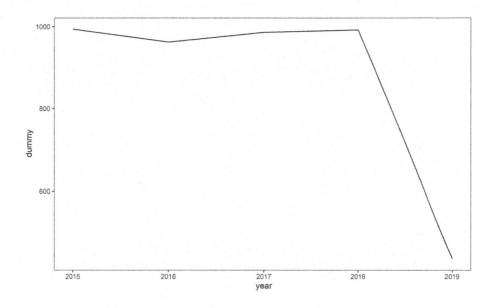

```
ggplot(yearly_shootings, aes(
  x = year,
  y = dummy
)) +
  geom_line() +
  theme_excel()
```

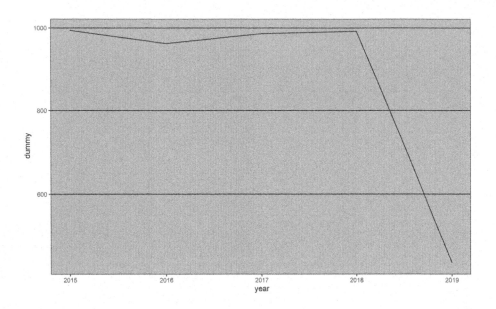

16

Hotspot maps

For this chapter you'll need the following file, which is available for download here[1]: san_francisco_suicide_2003_2017.csv.

Hotspot maps are used to find where events or places (e.g. crimes, marijuana dispensaries, liquors stores) are especially prevalent. These maps are frequently used by police departments, particularly in determining where to do hotspot policing (which focuses patrols on high-crime areas).

However, there are significant flaws with these kinds of maps. As we'll see during this lesson, minor changes to how we make the maps can cause significant differences in interpretation. For example, determining the size of the clusters that make up the hotspots can make it seem like there are much larger or smaller areas with hotspots than there actually are.

These clusters are also drawn fairly arbitrarily, without considering context such as neighborhoods (in Chapter 17 we'll make maps that try to account for these types of areas). This makes it more difficult to interpret because even though maps give us the context of location, it can combine different areas in an arbitrary way. Hotspot maps also often turn into population maps, where the dots indicate where people live rather than where the risk of something is. For example, a street with several apartment buildings will likely have more crimes (and thus have more dots on a hotspot map) than a street with only single-family homes. Maybe this is because the apartment street really is more crime-ridden than the single-family home street, but it could simply be that places with more people have more events (e.g. crimes, suicides, etc.) even if they actually have a lower rate of these events than less populated places. So not knowing the context of an area can make hotspot maps very misleading. We'll explore these issues in more detail throughout the lesson but keep in mind these risks as you make your own hotspot maps.

Here, we will make hotspot maps using data on suicides in San Francisco between 2003 and 2017. First, we need to read the data, which is called "san_francisco_suicide_2003_2017.csv". We can name the object we make *suicide*.

---

[1] https://github.com/jacobkap/crimebythenumbers/tree/master/data

```
library(readr)
suicide <- read_csv("data/san_francisco_suicide_2003_2017.csv")
suicide <- as.data.frame(suicide)
```

This data contains information on each crime reported in San Francisco including the type of crime (in our case always suicide), a more detailed crime category, and a number of date and location variables. Please note that suicide is not actually a crime, even though it is included in the San Francisco Police Department's crime data. As shown in Chapter 12, which also used San Francisco crime data, there are a number of other non-crimes included such as "Fire Report," "Traffic Collision," and "Non-Criminal." This is a fairly common occurrence in "crime" data where it also includes non-crimes that the police generally respond to so it is important to carefully examine your data to see what is included. Simply summing up the rows as a measure of crime will generally overcount crimes.

The columns X and Y are our longitude and latitude columns, which we will use to map the data.

```
head(suicide)
#    IncidntNum Category                        Descript
# 1  180318931  SUICIDE ATTEMPTED SUICIDE BY STRANGULATION
# 2  180315501  SUICIDE     ATTEMPTED SUICIDE BY JUMPING
# 3  180295674  SUICIDE            SUICIDE BY LACERATION
# 4  180263659  SUICIDE                          SUICIDE
# 5  180235523  SUICIDE     ATTEMPTED SUICIDE BY INGESTION
# 6  180236515  SUICIDE          SUICIDE BY ASPHYXIATION
#    DayOfWeek       Date     Time PdDistrict Resolution
# 1     Monday 04/30/2018 06:30:00    TARAVAL       NONE
# 2   Saturday 04/28/2018 17:54:00   NORTHERN       NONE
# 3   Saturday 04/21/2018 12:20:00   RICHMOND       NONE
# 4    Tuesday 04/10/2018 05:13:00    CENTRAL       NONE
# 5     Friday 03/30/2018 09:15:00    TARAVAL       NONE
# 6   Thursday 03/29/2018 17:30:00   RICHMOND       NONE
#                      Address         X        Y
# 1     0 Block of BRUCE AV -122.4517 37.72218
# 2   700 Block of HAYES ST -122.4288 37.77620
# 3   3700 Block of CLAY ST -122.4546 37.78818
# 4     0 Block of DRUMM ST -122.3964 37.79414
# 5 0 Block of FAIRFIELD WY -122.4632 37.72679
# 6   300 Block of 29TH AV -122.4893 37.78274
#                                         Location
# 1  POINT (-122.45168059935614 37.72218061554315)
```

```
# 2   POINT (-122.42876060987851 37.77620120112792)
# 3    POINT (-122.45462091999406 37.7881754224736)
# 4   POINT (-122.39642194376758 37.79414474237039)
# 5   POINT (-122.46324153155875 37.72679184368551)
# 6  POINT (-122.48929119750689 37.782735835121265)
#              PdId year
# 1 1.803189e+13 2018
# 2 1.803155e+13 2018
# 3 1.802957e+13 2018
# 4 1.802637e+13 2018
# 5 1.802355e+13 2018
# 6 1.802365e+13 2018
```

16.1   A simple map

To make these maps we will use the package ggmap.

```
install.packages("ggmap")
```

```
library(ggmap)
# Google's Terms of Service: https://cloud.google.com/maps-platform/terms/.
# Please cite ggmap if you use it! See citation("ggmap") for details.
#
# Attaching package: 'ggmap'
# The following object is masked from 'package:cowplot':
#
#     theme_nothing
# The following object is masked from 'package:tidygeocoder':
#
#     geocode
```

We'll start by making the background to our map, showing San Francisco. We do so by using the get_map() function from ggmap, which gets a map background from a number of sources. We'll set the source to "stamen" since Google no longer allows us to get a map without creating an account. The first parameter in get_map() is simply coordinates for San Francisco's bounding box to ensure we get a map of the right spot. A bounding box is four coordinates that connect to make a rectangle, used for determining where in the world to show.

An easy way to find the four coordinates for a bounding box is to go to the site Bounding Box.[2] This site has a map of the world and a box on the screen. Move the box to the area you want the map of. You may need to resize the box to cover the area you want. Then in the section that says "Copy & Paste," change the dropdown box to "CSV." In the section to the right of this are the four numbers that make up the bounding box. You can copy those numbers into `get_map()`

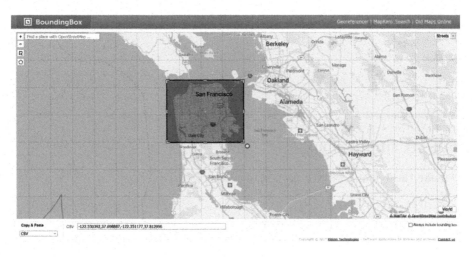

```
sf_map <- ggmap(get_map(c(-122.530392, 37.698887, -122.351177, 37.812996),
  source = "stamen"
))
sf_map
```

---

[2]https://boundingbox.klokantech.com/

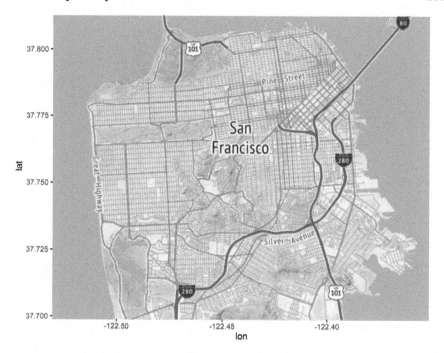

Since we saved the map output into *sf_map* we can reuse this map background for all the maps we're making in this lesson. This saves us time as we don't have to wait to download the map every time. Let's plot the suicides from our data set. Just as with a scatterplot we use the geom_point() function from the ggplot2 package and set our longitude and latitude variables on the x- and y-axis, respectively. When we load ggmap it also automatically loads ggplot2 as that package is necessary for ggmap to work, so we don't need to do library(ggplot2) ourselves.

```
sf_map +
  geom_point(aes(x = X, y = Y),
    data = suicide
  )
# Warning: Removed 1 rows containing missing values
# (geom_point).
```

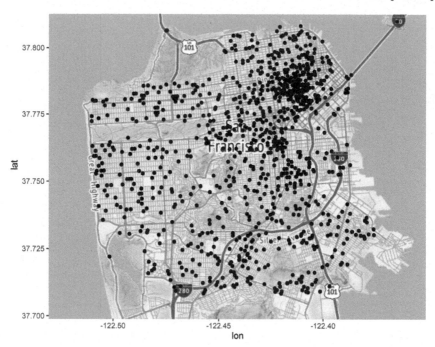

If we wanted to color the dots, we can use `color =` and then select a color. Let's try it with "forestgreen."

```
sf_map +
  geom_point(aes(x = X, y = Y),
    data  = suicide,
    color = "forestgreen"
  )
# Warning: Removed 1 rows containing missing values
# (geom_point).
```

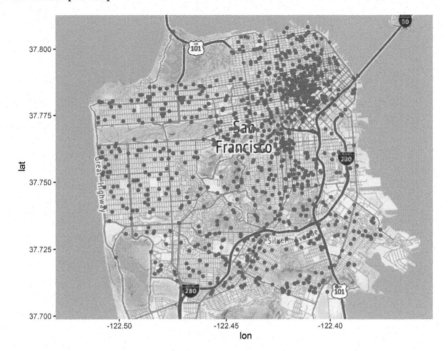

As with other graphs we can change the size of the dot using `size =`.

```
sf_map +
  geom_point(aes(x = X, y = Y),
    data  = suicide,
    color = "forestgreen",
    size  = 0.5
  )
# Warning: Removed 1 rows containing missing values
# (geom_point).
```

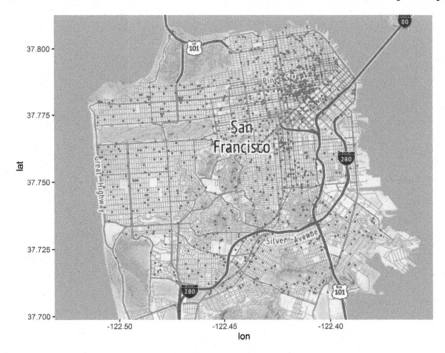

```
sf_map +
  geom_point(aes(x = X, y = Y),
    data  = suicide,
    color = "forestgreen",
    size  = 2
  )
# Warning: Removed 1 rows containing missing values
# (geom_point).
```

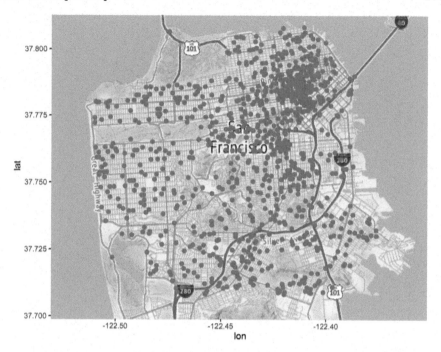

For maps like this - with one point per event - it is hard to tell if any events happen on the same, or nearly the same, location as each point is solid green. We want to make the dots semi-transparent so if multiple suicides happen at the same place that dot will be shaded darker than if only one suicide happened there. To do so we use the parameter `alpha =` which takes an input between 0 and 1 (inclusive). The lower the value the more transparent it is.

```
sf_map +
  geom_point(aes(x = X, y = Y),
    data  = suicide,
    color = "forestgreen",
    size  = 2,
    alpha = 0.5
  )
# Warning: Removed 1 rows containing missing values
# (geom_point).
```

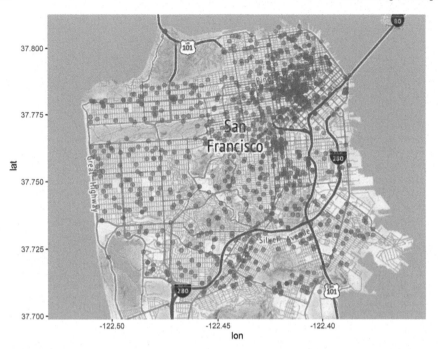

This map is useful because it allows us to easily see where each suicide in San Francisco happened between 2003 and 2017. There are some limitations though. For example, this shows all suicides in a single map, meaning that any time trends are lost.

16.2   What really are maps?

Let's pause for a moment to think about what a map really is. I made the following simple scatterplot of our data with one dot per suicide (minus the one without coordinates). Compare this to the previous map and you'll see that they are the same except the map has a useful background while the plot has a blank background. That is all static maps are (in Chapter 18 we'll learn about interactive maps), scatterplots of coordinates overlayed on a map background. Basically, they are scatterplots with context. And this context is useful; we can interpret the map to see that there are lots of suicides in the northeast part of San Francisco but not so many elsewhere, for example. The exact same pattern is present in the scatterplot but without the ability to tell "where" a dot is.

```
plot(suicide$X, suicide$Y, col = "forestgreen")
```

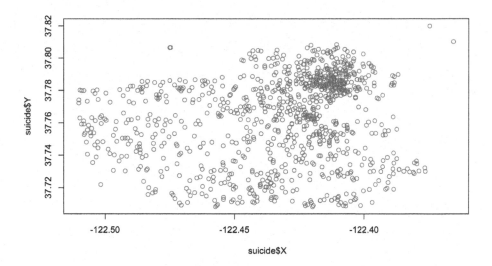

16.3 Making a hotspot map

Now we can start making hotspot maps, which help to show areas with clusters of events. We'll do this using hexagonal bins, which are an efficient way of showing clusters of events on a map. Our syntax will be similar to the map above, but now we want to use the function `stat_binhex()` rather than `geom_point()`. It starts the same as before with `aes(x = X, y = Y)` (or whatever the longitude and latitude columns are called in your data), as well as `data = suicide` outside of the `aes()` parameter.

There are two new things we need to make the hotspot map. First, we add the parameter `bins = number_of_bins` where "number_of_bins" is a number we select. `bins` essentially says how large or small we want each cluster of events to be. A smaller value for `bins` says we want more events clustered together, making larger bins. A larger value for bins has each bin be smaller on the map and capture fewer events. This will become clearer with examples.

The second thing is to add the function `coord_cartesian()`, which just tells `ggplot()` we are going to do some spatial analysis in the making of the bins. We don't need to add any parameters in this.

To use `stat_binhex()`, we'll also need to make sure that the package `hexbin` is installed. `stat_binhex()` will call the necessary function from `hexbin` internally so we don't need to run `library(hexbin)`.

```
install.packages("hexbin")
```

Let's start with 60 bins and then try some other number of bins to see how it changes the map.

```
sf_map +
  stat_binhex(aes(x = X, y = Y),
    bins = 60,
    data = suicide
  ) +
  coord_cartesian()
```

```
# Warning: Removed 1 rows containing non-finite values
# (stat_binhex).
```

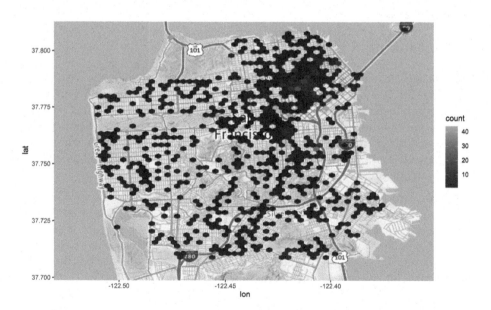

From this map we can see that most areas in the city had no suicides and that the areas with the most suicides are in downtown San Francisco.

What happens when we drop the number of bins to 30?

```
sf_map +
  stat_binhex(aes(x = X, y = Y),
    bins = 30,
    data = suicide
  ) +
  coord_cartesian()
```

```
# Warning: Removed 1 rows containing non-finite values
# (stat_binhex).
```

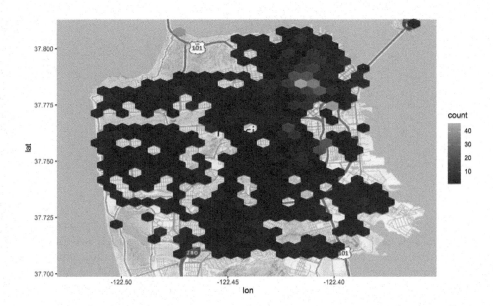

Each bin is much larger and covers nearly all of San Francisco. Be careful with maps like these! This map is so broad that it appears that suicides are ubiquitous across the city. We know from the map showing each suicide as a dot that there are fewer than 1,300 suicides; thus this is not true. Maps like this make it easy to mislead the reader, including yourself!

What about looking at 100 bins?

```
sf_map +
  stat_binhex(aes(x = X, y = Y),
    bins = 100,
    data = suicide
```

```
) +
coord_cartesian()
```

```
# Warning: Removed 1 rows containing non-finite values
# (stat_binhex).
```

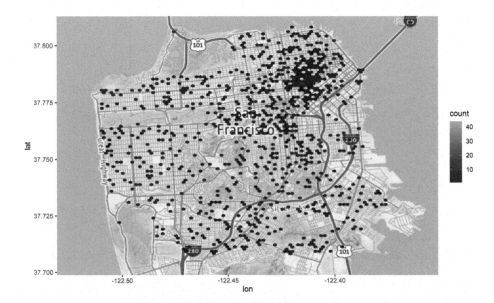

Now each bin is very small and a much smaller area in San Francisco has had a suicide. So what is the right number of bins to use? There is no correct universal answer - you must decide what the goal is with the data you are using. This opens up serious issues for manipulation - intentional or not - of the data as the map is so easily changeable without ever changing the data itself.

16.3.1 Colors

To change the bin colors we can use the parameter scale_fill_gradient(). This accepts a color for "low," which is when the events are rare, and "high" for the bins with frequent events. We'll use colors from ColorBrewer,[3] selecting the yellow-reddish theme ("3-class YlOrRd") from the Multi-hue section of the "sequential" data part of the page.

---

[3] http://colorbrewer2.org

```
sf_map +
  stat_binhex(aes(x = X, y = Y),
    bins = 60,
    data = suicide
  ) +
  coord_cartesian() +
  scale_fill_gradient(
    low = "#ffeda0",
    high = "#f03b20"
  )
```

```
# Warning: Removed 1 rows containing non-finite values
# (stat_binhex).
```

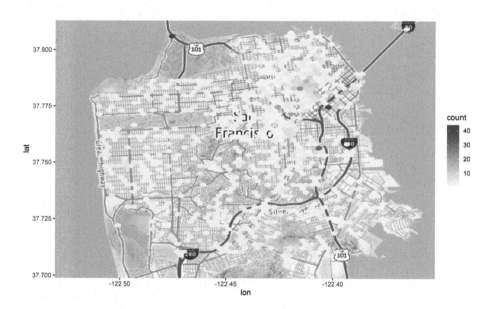

By default it labels the legend as "count." Since we know these are counts of suicides let's relabel that as such.

```
sf_map +
  stat_binhex(aes(x = X, y = Y),
    bins = 60,
    data = suicide
  ) +
  coord_cartesian() +
```

```
scale_fill_gradient("Suicides",
  low = "#ffeda0",
  high = "#f03b20"
)
```

```
# Warning: Removed 1 rows containing non-finite values
# (stat_binhex).
```

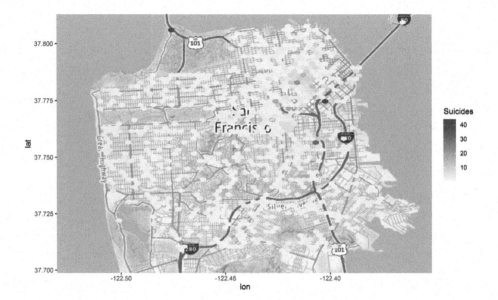

17

Choropleth maps

For this chapter you'll need the following files, which are available for download here[1]: san_francisco_suicide_2003_2017.csv, san_francisco_neighborhoods.dbf, san_francisco_neighborhoods.prj, san_francisco_neighborhoods.shp, san_francisco_neighborhoods.shx.

In Chapter 16 we made hotspot maps to show which areas in San Francisco had the most suicides. We made the maps in a number of ways and consistently found that suicides were most prevalent in northeast San Francisco. In this chapter we will make choropleth maps, which are shaded maps where each "unit" is some known geographic area, such as a state or neighborhood. Think of election maps where states are colored blue when a Democratic candidate wins that state and red when a Republican candidate wins. These are choropleth maps - each state is colored to indicate something. In this chapter we will continue to work on the suicide data and make choropleth maps shaded by the number of suicides in each neighborhood (we will define this later in the chapter) in the city.

Since we will be working more on the suicide data from San Francisco, let's read it in now.

```
library(readr)
suicide <- read_csv("data/san_francisco_suicide_2003_2017.csv")
suicide <- as.data.frame(suicide)
```

The package that we will use to handle geographic data and do most of the work in this chapter is sf. sf is a sophisticated package and does far more than what we will cover in this chapter. For more information about the package's features please see the website for it here.[2]

```
install.packages("sf")
```

---

[1] https://github.com/jacobkap/crimebythenumbers/tree/master/data
[2] http://r-spatial.github.io/sf/

```
library(sf)
```

For this chapter we will need to read in a shapefile that depicts the boundaries of each neighborhood in San Francisco. A shapefile is similar to a data.frame but has information on how to draw a geographic boundary such as a state. The way sf reads in the shapefiles is through the st_read() function. Our input inside the () is a string with the name of the ".shp" file we want to read in (since we are telling R to read a file on the computer rather than an object that exists, it needs to be in quotes). This shapefile contains neighborhoods in San Francisco so we'll call the object *sf_neighborhoods*.

I downloaded this data from San Francisco's Open Data site here,[3] selecting the Shapefile format in the Export tab. If you do so yourself it'll give you a zip file with multiple files in there. This is normal with shapefiles, you will have multiple files and only read in the file with the ".shp" extension to R. We still **do** need all of the files, and st_read() is using them even if not explicitly called. So make sure every file downloaded is in the same working directory as the .shp file. The files from this site had hard-to-read file names, so I relabeled them all as "san_francisco_neighborhoods" though that doesn't matter once it's read into R.

```
sf_neighborhoods <- st_read("data/san_francisco_neighborhoods.shp",
  quiet = TRUE
)
```

As usual when dealing with a new data set, let's look at the first 6 rows.

```
head(sf_neighborhoods)
# Simple feature collection with 6 features and 1 field
# Geometry type: MULTIPOLYGON
# Dimension:     XY
# Bounding box:  xmin: -122.4543 ymin: 37.70822 xmax: -122.357 ymax: 37.80602
# Geodetic CRS:  WGS84(DD)
#                          nhood
# 1        Bayview Hunters Point
# 2              Bernal Heights
# 3        Castro/Upper Market
# 4                   Chinatown
# 5                   Excelsior
```

---

[3]https://data.sfgov.org/Geographic-Locations-and-Boundaries/Analysis-Neighborhoods/p5b7-5n3h

```
# 6 Financial District/South Beach
#                            geometry
# 1 MULTIPOLYGON (((-122.3816 3...
# 2 MULTIPOLYGON (((-122.4036 3...
# 3 MULTIPOLYGON (((-122.4266 3...
# 4 MULTIPOLYGON (((-122.4062 3...
# 5 MULTIPOLYGON (((-122.424 37...
# 6 MULTIPOLYGON (((-122.3875 3...
```

The last column is important. In shapefiles, the "geometry" column is the one with the instructions to make the map. This data has a single row for each neighborhood in the city. So the "geometry" column in each row has a list of coordinates, which, if connected in order, make up that neighborhood. Since the "geometry" column contains the instructions to map, we can plot() it to show a map of the data.

```
plot(sf_neighborhoods$geometry)
```

Here we have a map of San Francisco broken up into neighborhoods. Is this a perfect representation of the neighborhoods in San Francisco? No. It is simply the city's attempt to create definitions of neighborhoods. Indeed, you're likely to find that areas at the border of neighborhoods are more similar to each other than they are to areas at the opposite side of their designated neighborhood. You can read a bit about how San Francisco determined the neighborhood boundaries here,[4] but know that this, like all geographic areas that someone has designated, has some degree of inaccuracy and arbitrariness in it. Like many things in criminology, this is just another limitation we will have to keep in mind.

In the `head()` results there was a section about something called "epsg" and "proj4string." Let's talk about that specifically since they are important for working with spatial data.

An issue with working with geographic data is that the Earth is not flat.[5] Since the Earth is spherical, there will always be some distortion when trying to plot the data on a flat surface such as a map. To account for this, we need to transform the longitude and latitude values we have to work properly on a map. We do so by "projecting" our data onto the areas of the Earth we want. This is a complex field with lots of work done on it (both abstractly and for R specifically) so this chapter will be an extremely brief overview of the topic and oversimplify some aspects of it.

If we look at the output of `st_crs(sf_neighborhoods)` we can see that the EPSG is set to 4326 and the proj4string (which tells us the current map projection) is "+proj=longlat +datum=WGS84 +no_defs". This CRS, WGS84, is a standard CRS and is the one used whenever you use a GPS to find a location. To find the CRS for certain parts of the world see here.[6] If you search that site for "California," you'll see that California is broken into 6 zones. The site isn't that helpful on which zones are which, but some Googling can often find state or region maps with the zones depicted there. We want California zone 3, which has the EPSG code 2227. We'll use this code to project this data properly.

If we want to get the proj4string for 2227 we can run `st_crs(2227)`. I'm not running it here because it will print out a large amount of text, but you should run it on your own computer. Note the text in text in this output includes "US survey foot." This means that the units are in feet. Some projections have units in meters so be mindful of this when doing some analysis, such as seeing if a point is within X feet of a certain area.

Let's convert our sf_neighborhoods data to coordinate reference system 2227 using `st_transform()`.

```
sf_neighborhoods <- st_transform(sf_neighborhoods, crs = 2227)
st_crs(sf_neighborhoods)
```

---

[4]https://data.sfgov.org/Geographic-Locations-and-Boundaries/Analysis-Neighborhoods/p5b7-5n3h
[5]https://en.wikipedia.org/wiki/Spherical_Earth
[6]https://spatialreference.org/

```
sf_neighborhoods <- st_transform(sf_neighborhoods, crs = 2227)
st_crs(sf_neighborhoods)
Coordinate Reference System:
  User input: EPSG:2227
  wkt:
PROJCRS["NAD83 / California zone 3 (ftUS)",
    BASEGEOGCRS["NAD83",
        DATUM["North American Datum 1983",
            ELLIPSOID["GRS 1980",6378137,298.257222101,
                LENGTHUNIT["metre",1]]],
        PRIMEM["Greenwich",0,
            ANGLEUNIT["degree",0.0174532925199433]],
        ID["EPSG",4269]],
    CONVERSION["SPCS83 California zone 3 (US Survey feet)",
        METHOD["Lambert Conic Conformal (2SP)",
            ID["EPSG",9802]],
        PARAMETER["Latitude of false origin",36.5,
            ANGLEUNIT["degree",0.0174532925199433],
            ID["EPSG",8821]],
        PARAMETER["Longitude of false origin",-120.5,
            ANGLEUNIT["degree",0.0174532925199433],
            ID["EPSG",8822]],
        PARAMETER["Latitude of 1st standard parallel",38.4333333333333,
            ANGLEUNIT["degree",0.0174532925199433],
            ID["EPSG",8823]],
        PARAMETER["Latitude of 2nd standard parallel",37.0666666666667,
            ANGLEUNIT["degree",0.0174532925199433],
            ID["EPSG",8824]],
        PARAMETER["Easting at false origin",6561666.667,
            LENGTHUNIT["US survey foot",0.304800609601219],
            ID["EPSG",8826]],
        PARAMETER["Northing at false origin",1640416.667,
            LENGTHUNIT["US survey foot",0.304800609601219],
            ID["EPSG",8827]]],
    CS[Cartesian,2],
        AXIS["easting (X)",east,
            ORDER[1],
            LENGTHUNIT["US survey foot",0.304800609601219]],
        AXIS["northing (Y)",north,
            ORDER[2],
            LENGTHUNIT["US survey foot",0.304800609601219]],
    USAGE[
        SCOPE["Engineering survey, topographic mapping."],
        AREA["United States (USA) - California - counties Alameda; Calaveras; Contra Cos
        BBOX[36.73,-123.02,38.71,-117.83]],
    ID["EPSG",2227]]
```

17.1   Spatial joins

What we want to do with these neighborhoods is to find out which neighborhood each suicide occurred in and sum up the number of suicides per neighborhood. Once we do that, we can make a map at the neighborhood level and be able to measure suicides per neighborhood. A spatial join is very similar to regular joins where we merge two data sets based on common variables (such as state name or unique ID code of a person). In this case it merges based on some shared geographic feature such as if two lines intersect or (as we will do so here) if a point is within some geographic area.

Right now our *suicide* data is in a data.frame with some info on each suicide and the longitude and latitude of the suicide in separate columns. We want to turn this data.frame into a spatial object to allow us to find which neighborhood each suicide happened in. We can convert it into a spatial object using the `st_as_sf()` function from `sf`. Our input is first our data, *suicide*. Then in the `coords` parameter we put a vector of the column names so the function knows which columns the longitude and latitude columns are so it can convert those columns to a "geometry" column like we saw in *sf_neighborhoods* earlier. We'll set the CRS to be the WGS84 standard we saw earlier, but we will change it to match the CRS that the neighborhood data has.

```
suicide <- st_as_sf(suicide,
  coords = c("X", "Y"),
  crs = "+proj=longlat +ellps=WGS84 +no_defs"
)
```

We want our suicides data in the same projection as the neighborhoods data so we need to use `st_transform()` to change the projection. Since we want the CRS to be the same as in *sf_neighborhoods*, we can set it using `st_crs(sf_neighborhoods)` to use the right CRS.

```
suicide <- st_transform(suicide,
  crs = st_crs(sf_neighborhoods)
)
```

Now we can take a look at `head()` to see if it was projected.

```
head(suicide)
# Simple feature collection with 6 features and 12 fields
# Geometry type: POINT
# Dimension:     XY
# Bounding box:  xmin: -122.4893 ymin: 37.72218 xmax: -122.3964 ymax: 37.79414
# Geodetic CRS:  WGS84(DD)
#    IncidntNum Category                          Descript
# 1  180318931  SUICIDE ATTEMPTED SUICIDE BY STRANGULATION
# 2  180315501  SUICIDE      ATTEMPTED SUICIDE BY JUMPING
# 3  180295674  SUICIDE            SUICIDE BY LACERATION
# 4  180263659  SUICIDE                          SUICIDE
# 5  180235523  SUICIDE      ATTEMPTED SUICIDE BY INGESTION
# 6  180236515  SUICIDE          SUICIDE BY ASPHYXIATION
#    DayOfWeek       Date     Time PdDistrict Resolution
# 1     Monday 04/30/2018 06:30:00    TARAVAL       NONE
# 2   Saturday 04/28/2018 17:54:00   NORTHERN       NONE
# 3   Saturday 04/21/2018 12:20:00   RICHMOND       NONE
# 4    Tuesday 04/10/2018 05:13:00    CENTRAL       NONE
# 5     Friday 03/30/2018 09:15:00    TARAVAL       NONE
# 6   Thursday 03/29/2018 17:30:00   RICHMOND       NONE
#                       Address
# 1      0 Block of BRUCE AV
# 2    700 Block of HAYES ST
# 3   3700 Block of CLAY ST
# 4      0 Block of DRUMM ST
# 5 0 Block of FAIRFIELD WY
# 6    300 Block of 29TH AV
#                                        Location
# 1   POINT (-122.45168059935614 37.72218061554315)
# 2   POINT (-122.42876060987851 37.77620120112792)
# 3    POINT (-122.45462091999406 37.7881754224736)
# 4   POINT (-122.39642194376758 37.79414474237039)
# 5   POINT (-122.46324153155875 37.72679184368551)
# 6 POINT (-122.48929119750689 37.782735835121265)
#            PdId year                  geometry
# 1 1.803189e+13 2018 POINT (-122.4517 37.72218)
# 2 1.803155e+13 2018  POINT (-122.4288 37.7762)
# 3 1.802957e+13 2018 POINT (-122.4546 37.78818)
# 4 1.802637e+13 2018 POINT (-122.3964 37.79414)
# 5 1.802355e+13 2018 POINT (-122.4632 37.72679)
# 6 1.802365e+13 2018 POINT (-122.4893 37.78274)
```

We can see it is now a "simple feature collection" with the correct projection. And we can see there is a new column called "geometry" just like in *sf_neighborhoods.* The type of data in "geometry" is POINT since our data is just a single location instead of a polygon like in the neighborhoods data.

Since we have both the neighborhoods and the suicides data let's make a quick map to see the data.

```
plot(sf_neighborhoods$geometry)
plot(suicide$geometry, add = TRUE, col = "red")
```

Our next step is to combine these two data sets to figure out how many suicides occurred in each neighborhood. This will be a multi-step process so let's plan it out before beginning. Our suicide data is one row for each suicide; our neighborhood data is one row for each neighborhood. Since our goal is to map at the neighborhood-level we need to get the neighborhood where each suicide occurred then aggregate up to the neighborhood-level to get a count of the suicides-per-neighborhood. Then we need to combine that with the original neighborhood data, and we can map it.

1. Find which neighborhood each suicide happened in
2. Aggregate suicide data until we get one row per neighborhood and a column showing the number of suicides in that neighborhood
3. Combine with the neighborhood data
4. Make a map

We'll start by finding the neighborhood where each suicide occurred using the function st_join(), which is a function in sf. This does a spatial join and finds the polygon (neighborhood in our case) where each point is located in. Since we will be aggregating the data, let's call the output of this function *suicide_agg*. The order in the () is important! For our aggregation we want the output to be at the suicide-level so we start with the *suicide* data. In the next step we'll see why this matters.

```
suicide_agg <- st_join(suicide, sf_neighborhoods)
```

Let's look at the first 6 rows.

```
head(suicide_agg)
# Simple feature collection with 6 features and 13 fields
# Geometry type: POINT
# Dimension:      XY
# Bounding box:  xmin: -122.4893 ymin: 37.72218 xmax: -122.3964 ymax: 37.79414
# Geodetic CRS:  WGS84(DD)
#   IncidntNum Category                         Descript
# 1  180318931  SUICIDE ATTEMPTED SUICIDE BY STRANGULATION
# 2  180315501  SUICIDE     ATTEMPTED SUICIDE BY JUMPING
# 3  180295674  SUICIDE            SUICIDE BY LACERATION
# 4  180263659  SUICIDE                          SUICIDE
# 5  180235523  SUICIDE     ATTEMPTED SUICIDE BY INGESTION
# 6  180236515  SUICIDE            SUICIDE BY ASPHYXIATION
#    DayOfWeek        Date      Time PdDistrict Resolution
# 1     Monday  04/30/2018 06:30:00    TARAVAL       NONE
# 2   Saturday  04/28/2018 17:54:00   NORTHERN       NONE
# 3   Saturday  04/21/2018 12:20:00   RICHMOND       NONE
# 4    Tuesday  04/10/2018 05:13:00    CENTRAL       NONE
# 5     Friday  03/30/2018 09:15:00    TARAVAL       NONE
# 6   Thursday  03/29/2018 17:30:00   RICHMOND       NONE
#                    Address
# 1     0 Block of BRUCE AV
# 2   700 Block of HAYES ST
# 3  3700 Block of CLAY ST
# 4     0 Block of DRUMM ST
# 5 0 Block of FAIRFIELD WY
# 6   300 Block of 29TH AV
#                                  Location
# 1   POINT (-122.45168059935614 37.72218061554315)
# 2   POINT (-122.42876060987851 37.77620120112792)
# 3   POINT (-122.45462091999406 37.7881754224736)
```

```
# 4  POINT (-122.39642194376758 37.79414474237039)
# 5  POINT (-122.46324153155875 37.72679184368551)
# 6 POINT (-122.48929119750689 37.782735835121265)
#            PdId year                             nhood
# 1 1.803189e+13 2018     Oceanview/Merced/Ingleside
# 2 1.803155e+13 2018                  Hayes Valley
# 3 1.802957e+13 2018               Presidio Heights
# 4 1.802637e+13 2018 Financial District/South Beach
# 5 1.802355e+13 2018             West of Twin Peaks
# 6 1.802365e+13 2018                Outer Richmond
#                       geometry
# 1 POINT (-122.4517 37.72218)
# 2  POINT (-122.4288 37.7762)
# 3 POINT (-122.4546 37.78818)
# 4 POINT (-122.3964 37.79414)
# 5 POINT (-122.4632 37.72679)
# 6 POINT (-122.4893 37.78274)
```

There is now the *nhood* column from the neighborhoods data, which says which neighborhood the suicide happened in. Now we can aggregate up to the neighborhood-level using `group_by()` and `summarize()` functions from the `dplyr` package.

We actually don't have a variable with the number of suicides so we need to make that. We can simply call it *number_suicides* and give it the value of 1 since each row is only one suicide.

```
suicide_agg$number_suicides <- 1
```

Now we can aggregate the data and assign the results back into *suicide_agg*.

```
library(dplyr)
suicide_agg <- suicide_agg %>%
  group_by(nhood) %>%
  summarize(number_suicides = sum(number_suicides))
```

Let's check a summary of the *number_suicides* variable we made.

```
summary(suicide_agg$number_suicides)
#    Min. 1st Qu.  Median    Mean 3rd Qu.    Max.
#    1.00   13.50   23.50   32.30   37.25  141.00
```

The minimum is one suicide per neighborhood, 32 on average, and 141 in the neighborhood with the most suicides. So what do we make of this data? Well, there are some data issues that cause problems in these results. Let's think about the minimum value. Did every single neighborhood in the city have at least one suicide? No. Take a look at the number of rows in this data, keeping in mind there should be one row per neighborhood.

```
nrow(suicide_agg)
# [1] 40
```

And let's compare it to the *sf_neighborhoods* data.

```
nrow(sf_neighborhoods)
# [1] 41
```

The suicides data is missing 2 neighborhoods (one of the 40 values is missing and is NA, not a real neighborhood). That is because if no suicides occurred there, there would never be a matching row in the data so that neighborhood wouldn't appear in the suicide data. That's not going to be a major issue here but is something to keep in mind if this were a real research project.

The data is ready to merge with the *sf_neighborhoods* data. We'll introduce a new function that makes merging data simple. This function also comes from the dplyr package.

The function we will use is left_join(), which takes two parameters, the two data sets to join together.

```
left_join(data1, data2)
```

This function joins these data and keeps all of the rows from the left data and every column from both data sets. It combines the data based on any matching columns (matching meaning same column name) in both data sets. Since in our data sets, the column *nhood* exists in both, it will merge the data based on that column.

There are two other functions that are similar but differ based on which rows they keep.

- left_join() - All rows from Left data and all columns from Left and Right data
- right_join() - All rows from Right data and all columns from Left and Right data
- full_join() - All rows and all columns from Left and Right data

We could alternatively use the merge() function, which is built into R, but that function is slower than the dplyr functions and requires us to manually set the matching columns.

We want to keep all rows in *sf_neighborhoods* (keep all neighborhoods) so we can use left_join(sf_neighborhoods, suicide_agg). Let's assign the results to a new data set called *sf_neighborhoods_suicide*.

We don't need the spatial data for "suicide_agg" anymore, and it will cause problems with our join if we keep it, so let's delete the "geometry" column from that data. We can do this by assigning the column the value of NULL.

```
suicide_agg$geometry <- NULL
```

Now we can do our join.

```
sf_neighborhoods_suicide <- left_join(sf_neighborhoods, suicide_agg)
# Joining, by = "nhood"
```

If we look at summary() again for *number_suicides* we can see that there are now two rows with NAs. These are the neighborhoods where there were no suicides so they weren't present in the *suicide_agg* data.

```
summary(sf_neighborhoods_suicide$number_suicides)
#    Min. 1st Qu.  Median    Mean 3rd Qu.    Max.    NA's
#    1.00   15.00   24.00   33.08   38.50  141.00       2
```

We need to convert these values to 0. We will use the is.na() function to conditionally find all rows with an NA value in the *number_suicides* column and use square bracket notation to change the value to 0.

```
sf_neighborhoods_suicide$number_suicides[
  is.na(sf_neighborhoods_suicide$number_suicides)
] <- 0
```

Checking it again we see that the minimum is now 0 and the mean number of suicides decreases a bit to about 31.5 per neighborhood.

```
summary(sf_neighborhoods_suicide$number_suicides)
#    Min. 1st Qu.  Median    Mean 3rd Qu.    Max.
#    0.00   12.00   23.00   31.46   36.00  141.00
```

17.2   Making choropleth maps

Finally we are ready to make some choropleth maps.

For these maps we are going to use ggplot2 again so we need to load it.

```
library(ggplot2)
```

ggplot2's benefit is you can slowly build graphs or maps and improve the graph at every step. Earlier, we used functions such as geom_line() for line graphs and geom_point() for scatter plots. For mapping these polygons we will use geom_sf(), which knows how to handle spatial data.

As usual we will start with ggplot(), inputting our data first. Then inside of aes (the aesthetics of the graph/map) we use a new parameter fill. In fill we will put in the *number_suicides* column, and it will color the polygons (neighborhoods) based on values in that column. Then we can add the geom_sf().

```
ggplot(sf_neighborhoods_suicide, aes(fill = number_suicides)) +
  geom_sf()
```

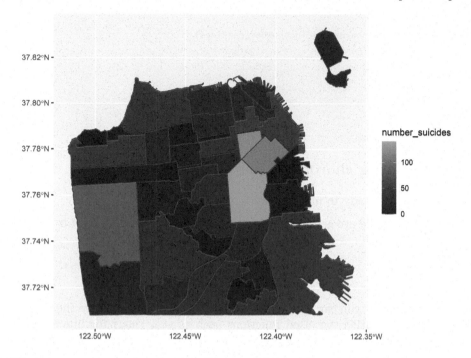

We have now created a choropleth map showing the number of suicides per neighborhood in San Francisco! Based on the legend, neighborhoods that are light blue have the most suicides while neighborhoods that are dark blue have the fewest (or none at all). Normally we'd want the opposite, with darker areas signifying a greater amount of whatever the map is showing.

We can use `scale_fill_gradient()` to set the colors to what we want. We input a color for low value and a color for high value, and it'll make the map shade by those colors.

```
ggplot(
  sf_neighborhoods_suicide,
  aes(fill = number_suicides)
) +
  geom_sf() +
  scale_fill_gradient(
    low = "white",
    high = "red"
  )
```

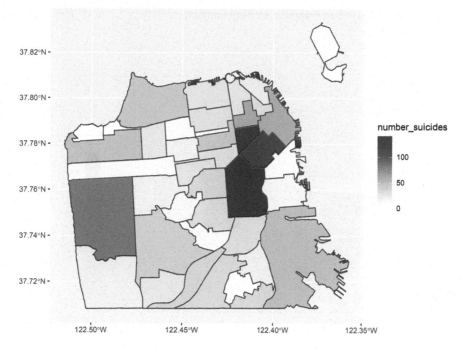

This gives a much better map and clearly shows the areas where suicides are most common and where there were no suicides.

To make this map easier to read and look better, let's add a title to the map and to the legend.

```
ggplot(
  sf_neighborhoods_suicide,
  aes(fill = number_suicides)
) +
  geom_sf() +
  scale_fill_gradient(
    low = "white",
    high = "red"
  ) +
  labs(
    fill = "# of suicides",
    title = "Suicides in San Francisco, by neighborhood",
    subtitle = "2003 - 2017"
  )
```

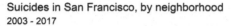

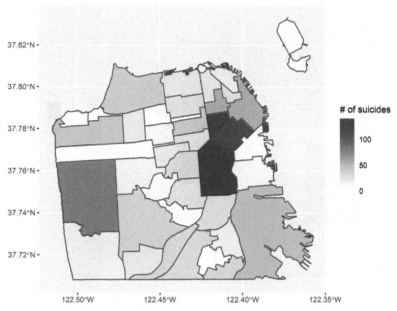

Since the coordinates don't add anything to the map, let's get rid of them.

```
ggplot(
  sf_neighborhoods_suicide,
  aes(fill = number_suicides)
) +
  geom_sf() +
  scale_fill_gradient(
    low = "white",
    high = "red"
  ) +
  labs(
    fill = "# of suicides",
    title = "Suicides in San Francisco, by neighborhood",
    subtitle = "2003 - 2017"
  ) +
  theme(
    axis.text.x = element_blank(),
    axis.text.y = element_blank(),
    axis.ticks = element_blank()
  )
```

Suicides in San Francisco, by neighborhood
2003 - 2017

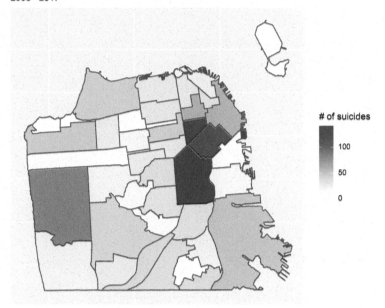

So what should we take away from this map? There are more suicides in the downtown area than any other place in the city. Does this mean that people are more likely to kill themselves there than elsewhere? Not necessarily. A major mistake people make when making a choropleth map (or really any type of map) is accidentally making a population map. The darker shaded parts of our map are also where a lot of people live. So if there are more people, it is reasonable that there would be more suicides (or crimes, etc.). What we'd really want to do is make a rate per some population (usually per 100k though this assumes equal risk for every person in the city which isn't really correct) to control for population differences.

We'll use this data in Chapter 18 to make interactive choropleth maps so let's save it.

```
save(sf_neighborhoods_suicide, file = "data/sf_neighborhoods_suicide.rda")
```

18

Interactive maps

For this chapter you'll need the following files, which are available for download here[1]: san_francisco_marijuana_geocoded.csv and sf_neighborhoods_suicide.rda.

While maps of data are useful, their ability to show incident-level information is quite limited. They tend to show broad trends - where crime happened in a city - rather than provide information about specific crime incidents. While broad trends are important, there are significant drawbacks about being unable to get important information about an incident without having to check the data. An interactive map bridges this gap by showing trends while allowing you to zoom into individual incidents and see information about each incident.

For this lesson we will be using data on every marijuana dispensary in San Francisco that has an active dispensary license as of late September 2019. The file is called "san_francisco_marijuana_geocoded.csv".

When downloaded from California's Bureau of Cannabis Control (here[2] if you're interested) the data contains the address of each dispensary but does not have coordinates. Without coordinates we are unable to map points, meaning we need to geocode them. Geocoding is the process of taking an address and getting the longitude and latitude of that address for mapping. For this lesson I've already geocoded the data, and we'll learn how to do so in Chapter 24.

```
library(readr)
marijuana <- read_csv("data/san_francisco_marijuana_geocoded.csv")
marijuana <- as.data.frame(marijuana)
```

---

[1]https://github.com/jacobkap/crimebythenumbers/tree/master/data
[2]https://aca5.accela.com/bcc/customization/bcc/cap/licenseSearch.aspx

18.1   Why do interactive graphs matter?

18.1.1   Understanding your data

The most important thing to learn from this book is that understanding your data is crucial to good research. Making interactive maps is a very useful way to better understand your data as you can immediately see geographic patterns and quickly look at characteristics of those incidents to understand them.

In this lesson we will make a map of each marijuana dispensary in San Francisco that lets you click on the dispensary and see some information about it. If we see a cluster of dispensaries, we can click on each one to see if they are similar - for example, if owned by the same person. Though it is possible to find these patterns just looking at the data, it is easier to be able to see a geographic pattern and immediately look at information about each incident.

18.1.2   Police departments use them

Interactive maps are popular in large police departments, such as Philadelphia and New York City. They allow easy understanding of geographic patterns in the data and, importantly, allow such access to people who do not have the technical skills necessary to interact with the data itself. If nothing else, learning interactive maps may help you with a future job.

18.2   Making the interactive map

As usual, let's take a look at the top 6 rows of the data.

```
head(marijuana)
#     License_Number                License_Type
# 1 C10-0000614-LIC Cannabis - Retailer License
# 2 C10-0000586-LIC Cannabis - Retailer License
# 3 C10-0000587-LIC Cannabis - Retailer License
# 4 C10-0000539-LIC Cannabis - Retailer License
# 5 C10-0000522-LIC Cannabis - Retailer License
# 6 C10-0000523-LIC Cannabis - Retailer License
```

```
#      Business_Owner         Business_Structure
# 1      Terry Muller Limited Liability Company
# 2    Jeremy Goodin                 Corporation
# 3     Justin Jarin                 Corporation
# 4 Ondyn Herschelle                 Corporation
# 5     Ryan Hudson Limited Liability Company
# 6     Ryan Hudson Limited Liability Company
#                        Premise_Address Status Issue_Date
# 1  2165 IRVING ST san francisco, CA 94122 Active   9/13/2019
# 2 122 10TH ST SAN FRANCISCO, CA 941032605 Active   8/26/2019
# 3   843 Howard ST SAN FRANCISCO, CA 94103 Active   8/26/2019
# 4    70 SECOND ST SAN FRANCISCO, CA 94105 Active    8/5/2019
# 5   527 Howard ST San Francisco, CA 94105 Active   7/29/2019
# 6 2414 Lombard ST San Francisco, CA 94123 Active   7/29/2019
#   Expiration_Date                 Activities
# 1       9/12/2020 N/A for this license type
# 2       8/25/2020 N/A for this license type
# 3       8/25/2020 N/A for this license type
# 4        8/4/2020 N/A for this license type
# 5       7/28/2020 N/A for this license type
# 6       7/28/2020 N/A for this license type
#   Adult-Use/Medicinal     lat       long
# 1                BOTH 37.76318 -122.4811
# 2                BOTH 37.77480 -122.4157
# 3                BOTH 37.78228 -122.4035
# 4                BOTH 37.78823 -122.4004
# 5                BOTH 37.78783 -122.3965
# 6                BOTH 37.79944 -122.4414
```

This data has information about the type of license, who the owner is, and where the dispensary is (as an address and as coordinates). We'll be making a map showing every dispensary in the city and make it so when you click a dot it'll make a popup showing information about that dispensary.

We will use the package leaflet for our interactive map. leaflet produces maps similar to Google Maps with circles (or any icon we choose) for each value we add to the map. It allows you to zoom in, scroll around, and provides context to each incident that isn't available on a static map.

```
install.packages("leaflet")
```

```
library(leaflet)
```

To make a `leaflet` map we need to run the function `leaflet()` and add a tile
to the map. We can just use the default tile which doesn't need an input. If
you're interested in other tiles, please see this website[3].

We will use a standard tile from Open Street Maps. This tile gives street
names and highlights important features such as parks and large stores which
provides useful contexts for looking at the data.

```
leaflet() %>%
  addTiles()
```

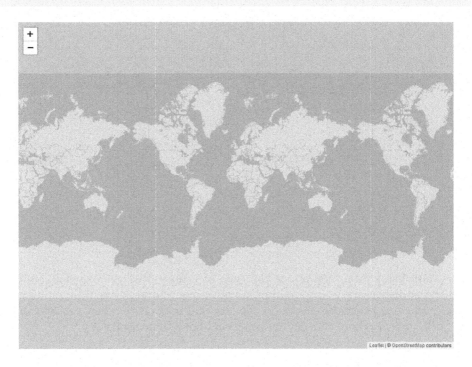

When you run the above code it shows a world map (copied several times).
Zoom into it, and it'll start showing relevant features of wherever you're looking.

Note the `%>%` between the `leaflet()` function and the `addTiles()` function. `leaflet`
is one of the packages in R where we can use pipes.

---

[3]https://leaflet-extras.github.io/leaflet-providers/preview/

To add the points to the graph we use the function `addMarkers()`, which has two parameters, `lng` and `lat`. For both parameters we put the column in which the longitude and latitude are, respectively.

```
leaflet() %>%
  addTiles() %>%
  addMarkers(
    lng = marijuana$long,
    lat = marijuana$lat
  )
```

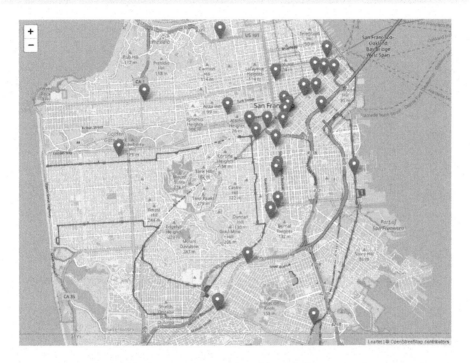

It now adds an icon indicating where every dispensary in our data is. You can zoom in and scroll around to see more about where the dispensaries are. There are only a few dozen locations in the data so the popups overlapping a bit doesn't affect our map too much. If we had more - such as crime data with millions of offenses - it would make it very hard to read. To change the icons to circles we can change the function `addMarkers()` to `addCircleMarkers()`, keeping the rest of the code the same.

```
leaflet() %>%
  addTiles() %>%
  addCircleMarkers(
    lng = marijuana$long,
    lat = marijuana$lat
  )
```

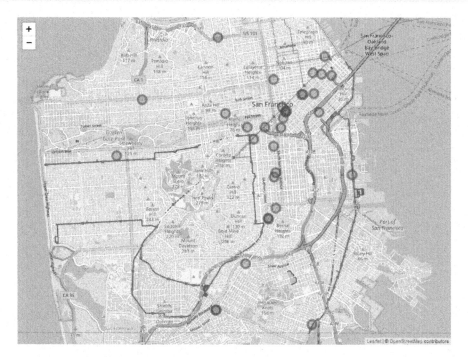

This makes the icon into circles, which take up less space than icons. To adjust the size of our icons we use the `radius` parameter in `addMarkers()` or `addCircleMarkers()`. The larger the radius, the larger the icons.

```
leaflet() %>%
  addTiles() %>%
  addCircleMarkers(
    lng = marijuana$long,
    lat = marijuana$lat,
    radius = 5
  )
```

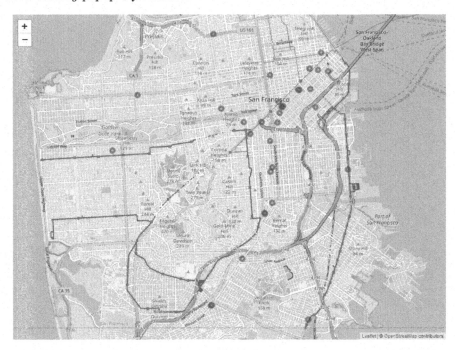

Setting the radius option to 5 shrinks the size of the icon a lot. In your own maps you'll have to fiddle with this option to get it to look the way you want. Let's move on to adding information about each icon when clicked upon.

18.3   Adding popup information

The parameter popup in the addMarkers() or addCircleMarkers() functions lets you input a character value (if not already a character value it will convert it to one) and that will be shown as a popup when you click on the icon. Let's start simple here by inputting the business owner column in our data and then build it up to a more complicated popup.

```
leaflet() %>%
  addTiles() %>%
  addCircleMarkers(
    lng = marijuana$long,
    lat = marijuana$lat,
    radius = 5,
```

```
    popup = marijuana$Business_Owner
)
```

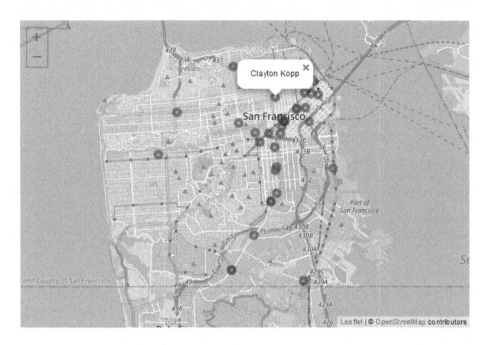

Try clicking around and you'll see that the owner of the dispensary you clicked on appears over the dot. If you're reading the print version of this book you won't, of course, be able to click on the map. We usually want to have a title indicating what the value in the popup means. We can do this by using the `paste()` function to combine text explaining the value with the value itself. Let's add the words "Business Owner:" before the business owner column.

```
leaflet() %>%
  addTiles() %>%
  addCircleMarkers(
    lng = marijuana$long,
    lat = marijuana$lat,
    radius = 5,
    popup = paste(
      "Business Owner:",
      marijuana$Business_Owner
    )
  )
```

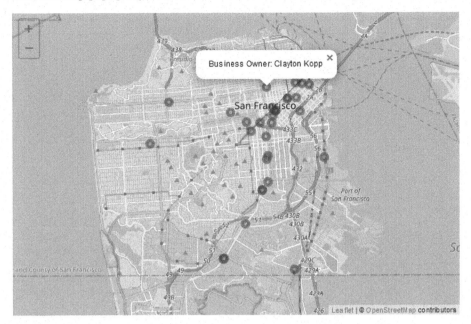

We don't have too much information in the data, but let's add the address and license number to the popup by adding them to the `paste()` function we're using.

```
leaflet() %>%
  addTiles() %>%
  addCircleMarkers(
    lng = marijuana$long,
    lat = marijuana$lat,
    radius = 5,
    popup = paste(
      "Business Owner:",
      marijuana$Business_Owner,
      "Address:",
      marijuana$Premise_Address,
      "License:",
      marijuana$License_Number
    )
  )
```

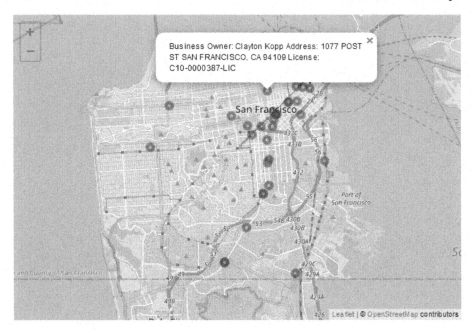

Just adding the location text makes it try to print out everything on one line, which is hard to read. If we add the text <br> where we want a line break, it will make one.
 is the HTML tag for line-break, which is why it works making a new line in this case.

```
leaflet() %>%
  addTiles() %>%
  addCircleMarkers(
    lng = marijuana$long,
    lat = marijuana$lat,
    radius = 5,
    popup = paste(
      "Business Owner:",
      marijuana$Business_Owner,
      "<br>",
      "Address:",
      marijuana$Premise_Address,
      "<br>",
      "License:",
      marijuana$License_Number
    )
  )
```

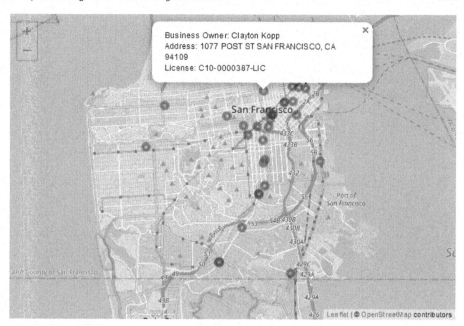

18.4 Dealing with too many markers

In our case with only 33 rows of data, turning the markers to circles solves our visibility issue. In cases with many more rows of data, this doesn't always work. A solution for this is to cluster the data into groups where the dots only show if you zoom in.

If we add the code `clusterOptions = markerClusterOptions()` to our `addCircleMark-ers()` it will cluster for us.

```
leaflet() %>%
  addTiles() %>%
  addCircleMarkers(
    lng = marijuana$long,
    lat = marijuana$lat,
    radius = 5,
    popup = paste(
      "Business Owner:",
      marijuana$Business_Owner,
      "<br>",
```

```
     "Address:",
     marijuana$Premise_Address,
     "<br>",
     "License:",
     marijuana$License_Number
   ),
   clusterOptions = markerClusterOptions()
)
```

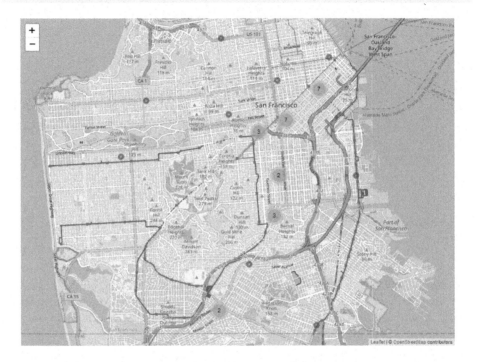

Locations close to each other are grouped together in fairly arbitrary group-ings, and we can see how large each grouping is by moving our cursor over the circle. Click on a circle or zoom in and it will show smaller groupings at lower levels of aggregation. Keep clicking or zooming in, and it will eventually show each location as its own circle.

This method is very useful for dealing with huge amounts of data as it avoids overflowing the map with too many icons at one time. A downside, however, is that the clusters are created arbitrarily meaning that important context, such as neighborhood, can be lost.

18.5   Interactive choropleth maps

In Chapter 17 we worked on choropleth maps which are maps with shaded regions, such as states colored by which political party won them in an election. Here we will make interactive choropleth maps where you can click on a shaded region and see information about that region. We'll make the same map as before - neighborhoods shaded by the number of suicides.

Let's load the San Francisco suicides-by-neighborhood data that we made earlier. We'll also want to project it to the standard longitude and latitude projection, otherwise our map won't work right.

```
library(sf)
# Linking to GEOS 3.9.1, GDAL 3.4.3, PROJ 7.2.1; sf_use_s2() is TRUE
load("data/sf_neighborhoods_suicide.rda")
sf_neighborhoods_suicide <- st_transform(
  sf_neighborhoods_suicide,
  "+proj=longlat +datum=WGS84"
)
```

We'll begin the leaflet map similar to before but use the function addPolygons(), and our input here is the geometry column of *sf_neighborhoods_suicide*.

```
leaflet() %>%
  addTiles() %>%
  addPolygons(data = sf_neighborhoods_suicide$geometry)
```

It made a map with thick blue lines indicating each neighborhood. Let's change the appearance of the graph a bit before making a popup or shading the neighborhoods The parameter `color` in `addPolygons()` changes the color of the lines - let's change it to black. The lines are also very thick, blurring into each other and making the neighborhoods hard to see. We can change the `weight` parameter to alter the size of these lines - smaller values are thinner lines. Let's try setting this to 1.

```
leaflet() %>%
  addTiles() %>%
  addPolygons(
    data = sf_neighborhoods_suicide$geometry,
    color = "black",
    weight = 1
  )
```

That looks better and we can clearly distinguish each neighborhood now.

As we did earlier, we can add the popup text directly to the function which makes the geographic shapes, in this case addPolygons(). Let's add the *nhood* column value - the name of that neighborhood - and the number of suicides that occurred in that neighborhood. As before, when we click on a neighborhood a popup appears with the output we specified.

```
leaflet() %>%
  addTiles() %>%
  addPolygons(
    data = sf_neighborhoods_suicide$geometry,
    col = "black",
    weight = 1,
    popup = paste0(
      "Neighborhood: ",
      sf_neighborhoods_suicide$nhood,
      "<br>",
      "Number of Suicides: ",
      sf_neighborhoods_suicide$number_suicides
    )
  )
```

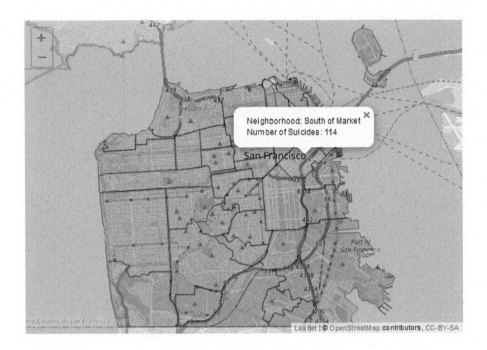

For these types of maps we generally want to shade each polygon to indicate how frequently the event occurred in the polygon. We'll use the function color-Numeric(), which takes a lot of the work out of the process of coloring in the map. This function takes two inputs, first a color palette, which we can get from the site Color Brewer.[4] Let's use the fourth bar in the Sequential page, which is light orange to red. If you look in the section with each HEX value it says that the palette is "3-class OrRd." The "3-class" just means we selected 3 colors, the "OrRd" is the part we want. That will tell colorNumeric() to make the palette using these colors. The second parameter is the column for our numeric variable, *number_suicides*. We will save the output of colorNumeric("OrRd", sf_neighborhoods_suicide$number_suicides) as a new object, which we'll call *pal* for convenience since it is a palette of colors. Then inside of addPolygons() we'll set the parameter fillColor to pal(sf_neighborhoods_suicide$number_suicides), running this function on the column. What this really does is determine which color every neighborhood should be based on the value in the *number_suicides* column.

```
pal <- colorNumeric("OrRd", sf_neighborhoods_suicide$number_suicides)
leaflet() %>%
  addTiles() %>%
```

---

[4] http://colorbrewer2.org/#type=sequential&scheme=OrRd&n=3

```
addPolygons(
  data = sf_neighborhoods_suicide$geometry,
  col = "black",
  weight = 1,
  popup = paste0(
    "Neighborhood: ",
    sf_neighborhoods_suicide$nhood,
    "<br>",
    "Number of Suicides: ",
    sf_neighborhoods_suicide$number_suicides
  ),
  fillColor = pal(sf_neighborhoods_suicide$number_suicides)
)
```

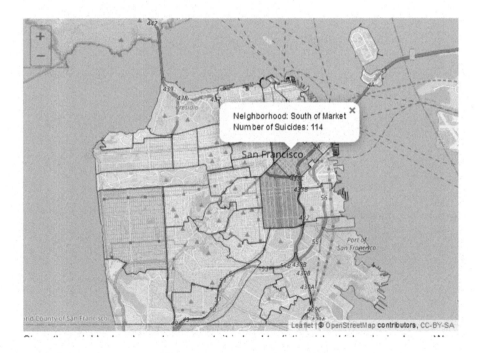

Since the neighborhoods are transparent, it is hard to distinguish which color is shown. We can make each neighborhood a solid color by setting the parameter fillOpacity inside of addPolygons() to 1.

```
leaflet() %>%
  addTiles() %>%
```

```
addPolygons(
  data = sf_neighborhoods_suicide$geometry,
  col = "black",
  weight = 1,
  popup = paste0(
    "Neighborhood: ",
    sf_neighborhoods_suicide$nhood,
    "<br>",
    "Number of Suicides: ",
    sf_neighborhoods_suicide$number_suicides
  ),
  fillColor = pal(sf_neighborhoods_suicide$number_suicides),
  fillOpacity = 1
)
```

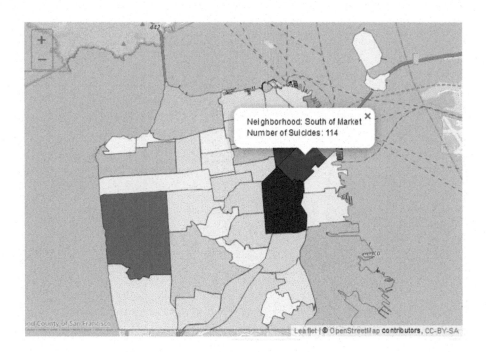

To add a legend to this we use the function `addLegend()`, which takes three parameters. `pal` asks which color palette we are using - we want it to be the exact same as we use to color the neighborhoods, so we'll use the *pal* object we made. The `values` parameter is used for which column our numeric values are from, in our case the *number_suicides* column so we'll input that. Finally `opacity` determines how transparent the legend will be. As each neighborhood is set to not be transparent at all, we'll also set this to 1 to be consistent.

```
leaflet() %>%
  addTiles() %>%
  addPolygons(
    data = sf_neighborhoods_suicide$geometry,
    col = "black",
    weight = 1,
    popup = paste0(
      "Neighborhood: ",
      sf_neighborhoods_suicide$nhood,
      "<br>",
      "Number of Suicides: ",
      sf_neighborhoods_suicide$number_suicides
    ),
    fillColor = pal(sf_neighborhoods_suicide$number_suicides),
    fillOpacity = 1
  ) %>%
  addLegend(
    pal = pal,
    values = sf_neighborhoods_suicide$number_suicides,
    opacity = 1
  )
```

Finally, we can add a title to the legend using the `title` parameter inside of `addLegend()`.

```
leaflet() %>%
  addTiles() %>%
  addPolygons(
    data = sf_neighborhoods_suicide$geometry,
    col = "black",
    weight = 1,
    popup = paste0(
      "Neighborhood: ",
      sf_neighborhoods_suicide$nhood,
      "<br>",
      "Number of Suicides: ",
      sf_neighborhoods_suicide$number_suicides
    ),
    fillColor = pal(sf_neighborhoods_suicide$number_suicides),
    fillOpacity = 1
  ) %>%
  addLegend(
    pal = pal,
    values = sf_neighborhoods_suicide$number_suicides,
    opacity = 1,
    title = "Suicides"
  ) %>%
  addProviderTiles(providers$CartoDB.Positron)
```

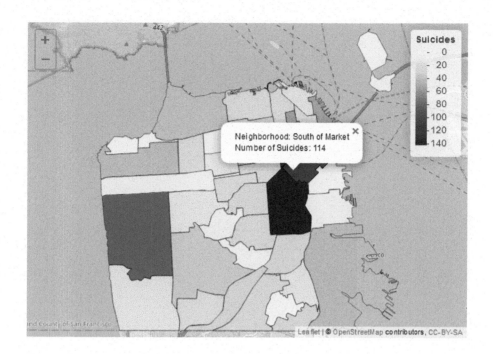

Part V

Collect

19

Webscraping with *rvest*

If I ever stop working in the field of criminology, I would certainly be a baker. So for the next few chapters we are going to work with "data" on baking. What we'll learn to do is find a recipe from the website All Recipes[1] and webscrape the ingredients and directions of that recipe.[2]

For our purposes we will be using the package rvest[3]. This package makes it relatively easy to scrape data from websites, especially when that data is already in a table on the page as our data will be.

If you haven't done so before, make sure to install rvest.

```
install.packages("rvest")
```

And every time you start R, if you want to use rvest you must tell R so by using library(rvest).

```
library(rvest)
#
# Attaching package: 'rvest'
# The following object is masked from 'package:readr':
#
#     guess_encoding
```

Here is a screenshot of the recipe for the "MMMMM... Brownies" (an excellent brownies recipe) page[4].

---

[1]https://www.allrecipes.com/

[2]The recipe was submitted by the user cicada77.

[3]https://github.com/tidyverse/rvest

[4]https://www.allrecipes.com/recipe/25080/mmmmm-brownies/?internalSource=hub%20recipe&referringContentType=Search

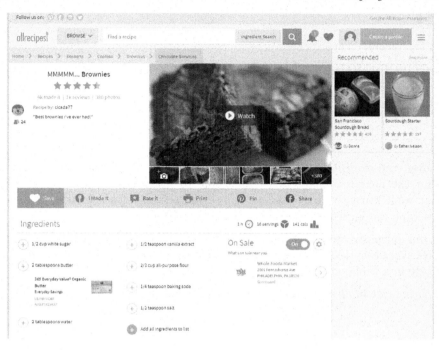

19.1 Scraping one page

In later lessons we'll learn how to scrape the ingredients of any recipe on the site. For now, we'll focus on just getting data for our brownies recipe.

The first step to scraping a page is to read in that page's information to R using the function `read_html()` from the `rvest` package. The input for the () is the URL of the page we want to scrape. In a later lesson, we will manipulate this URL to be able to scrape data from many pages.

```
# {html_document}
# <html id="recipeScTemplate_1-0" class="comp no-js taxlevel-5 recipeScTemplate
    html mntl-html" data-ab="60,99,62,99,84,99,99" data-resource-version
    ="1.11.0" lang="en" data-mantle-resource-version="3.13.509" data-allrecipes
    -resource-version="1.11.0" data-tracking-container="true">
# [1] <head class="loc head">\n<meta http-equiv="Content-Ty ...
# [2] <body>\n<svg class="mntl-svg-resource is-hidden"><def ...
```

```
read_html("https://www.allrecipes.com/recipe/25080/mmmmm-brownies/")
# {html_document}
# <html id="recipeScTemplate_1-0" class="comp no-js taxlevel-5 recipeScTemplate h
# [1] <head class="loc head">\n<meta http-equiv="Content-Ty ...
# [2] <body>\n<svg class="mntl-svg-resource is-hidden"><def ...
```

When running the above code, it returns an XML Document. The `rvest` package is well suited for interpreting this and turning it into something we already know how to work with. To be able to work on this data, we need to assign the output of `read_html()` to an object, which we'll call *brownies* since that is the recipe we are currently scraping.

```
brownies <- read_html("https://www.allrecipes.com/recipe/25080/mmmmm-brownies/")
```

We now need to select only a small part of the page that has the relevant information - in this case, the ingredients and directions.

We need to find just which parts of the page to scrape. To do so we'll use the helper tool SelectorGadget[5], a Google Chrome extension that lets you click

---

[5]https://selectorgadget.com/

on parts of the page to get the CSS selector code that we'll use. Install that extension in Chrome and go to the brownie recipe page.[6]

When you open SelectorGadget it allows you to click on parts of the page, and it will highlight every similar piece and show the CSS selector code in the box near the bottom. Here we clicked on the first ingredient - "1/2 cup white sugar." Every ingredient is highlighted in yellow as (to oversimplify this explanation) these ingredients are the same "type" in the page.

Note that in the bottom right of the screen, the SelectorGadget bar now has the text ".ingredients-item-name". This is the CSS selector code we can use to get all of the ingredients.

We will use the function `html_nodes()` to grab the part of the page (based on the CSS selectors) that we want. The input for this function is first the object made from `read_html()` (which we called *brownies*) and then we can paste the CSS selector text - in this case, ".ingredients-item-name". We'll assign the resulting object to *ingredients* since we want to use *brownies* to also get the directions.

---

[6] https://www.allrecipes.com/recipe/25080/mmmmm-brownies/?internalSource=hub %20recipe&referringContentType=Search

```
ingredients <- html_nodes(brownies, ".ingredients-item-name")
```

Since we are getting data that is a text format, we need to tell rvest that the format of the scraped data is text. We do with using html_text() and our input in the () is the object made in the function html_nodes().

```
ingredients <- html_text(ingredients)
```

Now let's check what we got.

```
ingredients
# character(0)
```

We have successfully scraped the ingredients for this brownies recipes.

Now let's do the same process to get the directions for baking.

In SelectorGadget click clear to unselect the ingredients. Now click one of the lines of directions that starts with the word "Step". It'll highlight all three directions as they're all of the same "type".[7] Note that if you click on the instructions without starting on one of the "Step" lines, such as clicking on the actual instructions (e.g. "Preheat the oven. . . ") lines itself, SelectorGadget will have the node "p" and say it has found 25 "things" on that page that match. To fix this you just scroll up to see where the text "Best brownies I've ever had!" is also highlighted in yellow and click that to unselect it. Using SelectorGadget is often steps like this where you use trial and error to only select the parts of the page that you want.

---

[7]To be slightly more specific, when the site is made it has to put all of the pieces of the site together, such as links, photos, the section on ingredients, the section on directions, the section on reviews. So in this case we selected a "text" type in the section on directions and SelectorGadget then selected all "text" types inside of that section.

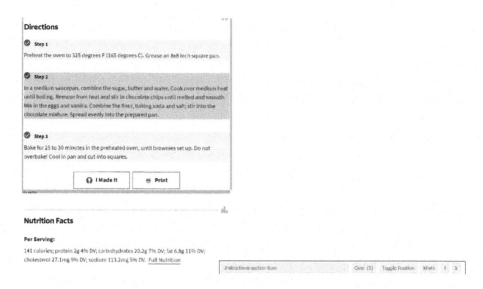

The CSS selector code this time is ".instructions-section-item" so we can put that inside of `html_nodes()`. Let's assign the output as *directions*.

```
directions <- html_nodes(brownies, ".instructions-section-item")
directions <- html_text(directions)
```

Did it work?

```
directions
```

```
directions
#> [1] "  Step 1   Preheat the oven to 325 degrees F (165 degrees C). Grease an 8x8 inch square p
#> [2] "  Step 2   In a medium saucepan, combine the sugar, butter and water. Cook over medium he
#> [3] "  Step 3   Bake for 25 to 30 minutes in the preheated oven, until brownies set up. Do not
```

Yes! You may notice that each direction is one very long string, so long that we have to scroll to the right (in the web version of this book) to read it. If you run the code direction in RStudio, it'll automatically put it on multiple lines for easy reading. If you put it on a website or a PDF, it'll instead be so long that it may extend off the page. There are many features in RStudio that make it easy to work with data like this. In cases where you are presenting the data outside of RStudio, such as making an R Markdown document, it is important to check that the results look right in every format you are making (e.g. Word, HTML, PDF).

19.2   Cleaning the webscraped data

Now we just need to clean up the extra spaces to have nice, clean instructions to make the brownies from the recipe we scraped. We can remove white space at the beginning or end of strings using the `trimws()` function that is built into R. We just put the vector object inside the parentheses.

```
directions <- trimws(directions)
ingredients <- trimws(ingredients)
```

And let's print out both objects to make sure it worked.

```
ingredients
directions
```

```
ingredients
#> [1] "% cup white sugar"
#> [2] "2 tablespoons butter"
#> [3] "2 tablespoons water"
#> [4] "1 % cups semisweet chocolate chips"
#> [5] "2 eggs"
#> [6] "% teaspoon vanilla extract"
#> [7] "<U+2154> cup all-purpose flour"
#> [8] "% teaspoon baking soda"
#> [9] "% teaspoon salt"
directions
#> [1] "Step 1   Preheat the oven to 325 degrees F (165 degrees C). Grease an 8x8 inch square pan.
#> [2] "Step 2   In a medium saucepan, combine the sugar, butter and water. Cook over medium heat
#> [3] "Step 3   Bake for 25 to 30 minutes in the preheated oven, until brownies set up. Do not ov
```

Now *ingredients* is as it should be, though note that all of the ingredient amounts - e.g. 2/3 cups - looks fine when in R. But when exporting it to PDF or HTML it shows weird characters like "<U+2154>." This is because the conversion from R to PDF or HTML isn't working right. I'm keeping this unfixed as a demonstration of how things can look right in R but look wrong when moving it elsewhere. So when working on something that you export out of R (including from R to PDF/HTML or even R to Excel), you should make sure to check that no issue occurred during the conversion.

directions has a bunch of space between the step number and the instructions. Let's use gsub() to remove the multiple spaces and replace it with a single space.

We'll search for anything with two or more spaces and replace that with a single space.

```
directions <- gsub(" {2,}", " ", directions)
```

And one final check to make sure it worked.

```
directions
```

```
directions
#> [1] "Step 1 Preheat the oven to 325 degrees F (165 degrees C). Grease an 8x8 inch square pan. A
#> [2] "Step 2 In a medium saucepan, combine the sugar, butter and water. Cook over medium heat un
#> [3] "Step 3 Bake for 25 to 30 minutes in the preheated oven, until brownies set up. Do not over
```

In Chapter 20 we'll learn to make a function to scrape any recipe from this site using just the URL and to print the ingredients and directions to the console.

20

Functions

So far, we have been writing code to handle specific situations, such as subsetting a single data set, often using other people's functions. In cases where you want to reuse the code, it is unwise to simply copy and paste the code and make minor changes to handle the new data. Instead we want something that is able to take multiple values and perform the same action (subset, aggregate, make a plot, webscrape, etc.) on those values. We've used lots of other people's function throughout this book, and in this chapter we'll learn how to create our own.

Think of a function like a stapler - you put the paper in, push down, and it staples the paper together. It doesn't matter what papers you are using; it always staples them together. If you needed to buy a new stapler every time you needed to staple something (i.e. copy and pasting code) you'd quickly have way too many staplers (and waste a bunch of money).

An important benefit is that you can use this function again and again to help solve other problems. Let's imagine you need to clean crime data from 10 different cities. Most cities' crime data is very similar so writing the code for one gets you most of the way there for the other 9 cities. Your code will probably work for the other cities, with only minor changes necessary (for example, column names are probably different across all agencies). However, copy and pasting code quickly becomes a terrible solution - functions work much better. If you did copy and paste 10 times and you found a bug, then you'd have to fix the bug 10 times. With a function, you would change the code once.

20.1 A simple function

We'll start with a simple function that takes a number and returns that number plus the number 2.

```
add_2 <- function(number) {
  number <- number + 2
  return(number)
}
```

The syntax (how we write it) of a function is

function_name <- function(parameters) {
code
return(output)
}

There are five essential parts of a function

- function_name - This is just the name we give to the function. It can be anything, but as when making other objects, call it something that is easy to remember what it does.
- parameters - Here is where we say what goes into the function. In most cases you will want to put some data in and expect something new out. For example, for the function mean() you put in a vector of numbers in the () section, and it returns the mean of those numbers. Here is also where you can put any options to affect how the code is run.
- code - This is the code you write to do the thing you want the function to do. In the above example our code is number <- number + 2. For any number inputted, our code adds 2 to it and assigns it back into the object *number*.
- return - This is something new in this book, here you use the return() function and inside the () you put the object you want to be outputted. In our example we have *number* inside the return() as that's what we want to come out of the function. It is not always necessary to end your function with return() but is highly recommended to do so to make sure you're outputting what it is you want to output.
- The final piece is the structure of your function. After the function_name (whatever it is you call it) you always need the text <- function() where the parameters (if any) are in the (). After the closing parentheses put a {, and at the very end of the function, after the return(), close those squiggly brackets with a }. The <- function() tells R that you are making a function rather than some other type of object. And the { and } tell R that all the code in between are part of that function.

Our function here adds 2 to any number we input.

```
add_2(2)
# [1] 4
```

```
add_2(5)
# [1] 7
```

20.2 Adding parameters

Let's add a single parameter, which multiplies the result by 5 if selected.

```
add_2 <- function(number, times_5 = FALSE) {
  number <- number + 2
  return(number)
}
```

Now we have added a parameter called time_5 to the () part of the function and set it the be FALSE by default. Right now it doesn't do anything so we need to add code to say what happens if it is TRUE (remember in R true and false must always be all capital letters and not in quotes).

```
add_2 <- function(number, times_5 = FALSE) {
  number <- number + 2

  if (times_5 == TRUE) {
    number <- number * 5
  }

  return(number)
}
```

Now our code says if the parameter times_5 is TRUE, then do the thing in the squiggly brackets {} below. Note that we use the same squiggly brackets as when making the entire function. That just tells R that the code in those brackets belong together. Let's try out our function.

```
add_2(2)
# [1] 4
```

It returns 4, as expected. Since the parameter `times_5` is defaulted to FALSE, we don't need to specify that parameter if we want it to stay FALSE. When we don't tell the function that we want it to be TRUE, the code in our "if statement" doesn't run. When we set `times_5` to TRUE, it runs that code.

```
add_2(2, times_5 = TRUE)
# [1] 20
```

20.3   Making a function to scrape recipes

In Section 19.1 we wrote some code to scrape data from the website All Recipes[1] for a recipe. We are going to turn that code into a function here. The benefit is that our input to the function will be a URL, and then it will print out the ingredients and directions for that recipe. If we want multiple recipes (and for webscraping you usually will want to scrape multiple pages), we just change the URL we input without changing the code at all.

We used the `rvest` package so we need to tell R we want to use it again.

```
library(rvest)
```

Let's start by writing a shell of the function - everything but the code. We can call it *scrape_recipes* (though any name would work), add in the `<- function()` and put URL in the () as our input for the function is the URL of the page with the recipe we want. For this function we won't return any object, we will just print things to the console, so we don't need the `return()` value. Don't forget the `{` after the end of the `function()` and `}` at the very end of the function.

```
scrape_recipes <- function(URL) {

}
```

Now we need to add the code that takes the URL, scrapes the website, and assigns the ingredients part of the page to an object called *ingredients* and the directions part to an object called *directions*. Since we have the code from an

---

[1] https://www.allrecipes.com/

earlier lesson, we can copy and paste that code into the function and make a small change to get a working function.

```
scrape_recipes <- function(URL) {
  brownies <- read_html("https://www.allrecipes.com/recipe/25080/mmmmm-brownies/")

  ingredients <- html_nodes(brownies, ".ingredients-item-name")
  ingredients <- html_text(ingredients)

  directions <- html_nodes(brownies, ".instructions-section-item")
  directions <- html_text(directions)
  directions <- trimws(directions)
}
```

The part inside the () of read_html() is the URL of the page we want to scrape. This is the part of the function that will change based on our input. We want whatever input is in the URL parameter to be the URL we scrape. So let's change the URL of the brownies recipe we scraped previously to simply say URL (without quotes).

```
scrape_recipes <- function(URL) {
  brownies <- read_html(URL)

  ingredients <- html_nodes(brownies, ".ingredients-item-name")
  ingredients <- html_text(ingredients)

  directions <- html_nodes(brownies, ".instructions-section-item")
  directions <- html_text(directions)
  directions <- trimws(directions)
}
```

To make this function print something to the console, we need to specifically tell it to do so in the code. We do this using the print() function. Let's first print the ingredients and then the directions. We'll add that to the final lines of the function.

```
scrape_recipes <- function(URL) {
  brownies <- read_html(URL)

  ingredients <- html_nodes(brownies, ".ingredients-item-name")
  ingredients <- html_text(ingredients)
```

```
directions <- html_nodes(brownies, ".instructions-section-item")
directions <- html_text(directions)
directions <- trimws(directions)

print(ingredients)
print(directions)
}
```

Now we can try it for a new recipe, this one for "The Best Lemon Bars" at this link[2].

```
scrape_recipes("https://www.allrecipes.com/recipe/10294/the-best-lemon-bars/")
```

```
scrape_recipes("https://www.allrecipes.com/recipe/10294/the-best-lemon-bars/")
#> [1] "1 cup butter, softened "     "¾ cup white sugar "
#> [3] "2 cups all-purpose flour " "4 eggs "
#> [5] "1 ½ cups white sugar "      "¼ cup all-purpose flour "
#> [7] "2 lemons, juiced "
#> [1] "Step 1   Preheat oven to 350 degrees F (175 degrees C).   Advertisement"
#> [2] "Step 2   In a medium bowl, blend together softened butter, 2 cups flour and 1/2 cup sugar.
#> [3] "Step 3   Bake for 15 to 20 minutes in the preheated oven, or until firm and golden. In ano
#> [4] "Step 4   Bake for an additional 20 minutes in the preheated oven. The bars will firm up as
```

In the next lesson we'll use "for loops" to scrape multiple recipes very quickly.

---

[2]https://www.allrecipes.com/recipe/10294/the-best-lemon-bars/

21

For loops

We will often want to perform the same task on a number of different items, such as cleaning every column in a data set. One effective way to do this is through "for loops". Earlier in this book we learned how to scrape the recipe website All Recipes.[1] We did so for a single recipe. If we wanted to get a feast's worth of recipes, typing out each recipe would be slow, even with the function we made in Section 20.3. In this chapter we will use a for loop to scrape multiple recipes very quickly.

21.1 Basic for loops

We'll start with a simple example of a for loop, making R print the numbers 1-10.

```
for (i in 1:10) {
  print(i)
}
# [1] 1
# [1] 2
# [1] 3
# [1] 4
# [1] 5
# [1] 6
# [1] 7
# [1] 8
# [1] 9
# [1] 10
```

The basic concept of a for loop is that you have some code that you need to run many times with slight changes to a value or values in the code - somewhat like

---
[1] https://www.allrecipes.com/

a function. And like a function, all the code you want to use goes in between the { and } squiggly brackets. And you loop through all the values you specify - meaning that the code runs once for each of those values.

Let's look closer at the (i in 1:10). The i is simply a placeholder object, which takes the value 1 through 10 each iteration of the loop. An iteration is the formal term for each time the loop runs. In our loop it will run 10 times as we have 10 numbers (1-10). The first time it runs the i gets the value of 1, the second time it runs i gets the value of 2, and so on.

It's not necessary to call it i, but it is the convention in programming to do so. It takes the value of whatever follows the in, which can range from a vector of strings or numbers to lists of data.frames (though we won't do anything that complicated in this chapter). Especially when you're an early learner of R, it could help to call the i something informative to you about what value it has. Let's go through a few examples with different names for i and different values it is looping through.

```r
for (a_number in 1:10) {
  print(a_number)
}
# [1] 1
# [1] 2
# [1] 3
# [1] 4
# [1] 5
# [1] 6
# [1] 7
# [1] 8
# [1] 9
# [1] 10
```

```r
animals <- c("cat", "dog", "gorilla", "buffalo", "lion", "snake")
for (animal in animals) {
  print(animal)
}
# [1] "cat"
# [1] "dog"
# [1] "gorilla"
# [1] "buffalo"
# [1] "lion"
# [1] "snake"
```

Now let's make our code a bit more complicated, adding the number 2 every loop.

```
for (a_number in 1:10) {
  print(a_number + 2)
}
# [1] 3
# [1] 4
# [1] 5
# [1] 6
# [1] 7
# [1] 8
# [1] 9
# [1] 10
# [1] 11
# [1] 12
```

We're keeping the results inside of `print()` since for loops do not print the results by default. Let's try combining this with some subsetting using square bracket notation `[]`. We will look through every value in *numbers*, a vector we will make with the values 1:10, and replace each value with its value plus 2.

The object we're looping through is *numbers*. But we're actually looping through every index it has, hence the `1:length(numbers)`. That is saying, `i` takes the value of each index in *numbers*, which is useful when we want to change that element. `length(numbers)` finds how long the vector *numbers* is (if this was a data.frame we could use `nrow()`) to find how many elements it has. In the code we take the value at each index `numbers[i]` and add 2 to it.

```
numbers <- 1:10
for (i in 1:length(numbers)) {
  numbers[i] <- numbers[i] + 2
}
```

```
numbers
# [1]  3  4  5  6  7  8  9 10 11 12
```

We can also include functions we made in for loops. Here's a function we made last chapter which adds 2 to each inputted number.

```
add_2 <- function(number) {
  number <- number + 2
  return(number)
}
```

Let's put that in the loop.

```
for (i in 1:length(numbers)) {
  numbers[i] <- add_2(numbers[i])
}
```

```
numbers
# [1]  5  6  7  8  9 10 11 12 13 14
```

21.2   Scraping multiple recipes

Below is the function copied from Section 20.3 which takes a single URL and scraped the site All Recipes[2] for that recipe. It printed the ingredients and directions to cook that recipe to the Console. If we wanted to get that info for multiple recipes, we would need to run the function multiple times. Here we will use a for loop to do this. Since we're using the read_html() function from rvest, we need to tell R we want to use that package.

```
library(rvest)
scrape_recipes <- function(URL) {
  brownies <- read_html(URL)

  ingredients <- html_nodes(brownies, ".ingredients-item-name")
  ingredients <- html_text(ingredients)

  directions <- html_nodes(brownies, ".instructions-section-item")
  directions <- html_text(directions)
  directions <- trimws(directions)
```

---

[2]https://www.allrecipes.com/

```
  print(ingredients)
  print(directions)
}
```

With any for loop you need to figure out what is going to be changing, in this case it is the URL. And since we want multiple recipes, we will make a vector with the URLs of all the recipes we want.

Here I am making a vector called *recipe_urls* with the URLs of a few recipes that I like on the site. The way I got the URLs was to go to each recipe's page and copy and paste the URL. Is this the right approach? Shouldn't we do everything in R? Not always. In situations like this where we know that there are a small number of links we want, it is reasonable to do it by hand. Remember that R is a tool to help you. While keeping everything you do in R is good for reproducibility, it is not always reasonable and may take too much time or effort given the constraints - usually limited time - of your project.

```
recipe_urls <- c(
  "https://www.allrecipes.com/recipe/25080/mmmmm-brownies/",
  "https://www.allrecipes.com/recipe/27188/crepes/",
  "https://www.allrecipes.com/recipe/22180/waffles-i/"
)
```

Now we can write the for loop to go through every single URL in *recipe_urls* and use the function `scrape_recipes` on that URL.

```
for (recipe_url in recipe_urls) {
  scrape_recipes(recipe_url)
}
```

```
for (recipe_url in recipe_urls) {
  scrape_recipes(recipe_url)
}
#> [1] "¼ cup white sugar "
#> [2] "2 tablespoons butter "
#> [3] "2 tablespoons water "
#> [4] "1 ¼ cups semisweet chocolate chips "
#> [5] "2 eggs "
#> [6] "½ teaspoon vanilla extract "
#> [7] "<U+2154> cup all-purpose flour "
#> [8] "¼ teaspoon baking soda "
#> [9] "¼ teaspoon salt "
#> [1] "Step 1   Preheat the oven to 325 degrees F (165 degrees C). Grease an 8x8 inch square pan.
#> [2] "Step 2   In a medium saucepan, combine the sugar, butter and water. Cook over medium heat
#> [3] "Step 3   Bake for 25 to 30 minutes in the preheated oven, until brownies set up. Do not ov
#> [1] "2 eggs "
#> [2] "1 cup milk "
#> [3] "<U+2154> cup all-purpose flour "
#> [4] "1 pinch salt "
#> [5] "1 ½ teaspoons vegetable oil "
#> [1] "Step 1   In a blender combine eggs, milk, flour, salt and oil. Process until smooth. Cover
#> [2] "Step 2   Heat a skillet over medium-high heat and brush with oil. Pour 1/4 cup of crepe ba
#> [1] "2 eggs "
#> [2] "2 cups all-purpose flour "
#> [3] "1 ¼ cups milk "
#> [4] "½ cup vegetable oil "
#> [5] "1 tablespoon white sugar "
#> [6] "4 teaspoons baking powder "
#> [7] "¼ teaspoon salt "
#> [8] "½ teaspoon vanilla extract "
#> [1] "Step 1   Preheat waffle iron. Beat eggs in large bowl with hand beater until fluffy. Beat
#> [2] "Step 2   Spray preheated waffle iron with non-stick cooking spray. Pour mix onto hot waffl
```

22

Scraping tables from PDFs

For this chapter you'll need the following file, which is available for download here[1]: usbp_stats_fy2017_sector_profile.pdf.

Government agencies in particular like to release their data in long PDFs which often have the data we want in a table on one of the pages. To use this data we need to scrape it from the PDF into R. In the majority of cases when you want data from a PDF it will be in a table. Essentially the data will be an Excel file inside of a PDF. This format is not altogether different from what we've done before.

Let's first take a look at the data we will be scraping. The first step in any PDF scraping should be to look at the PDF and try to think about the best way to approach this particular problem. While all PDF scraping follows a general format, you cannot necessarily reuse your old code as each situation is likely slightly different. Our data is from the US Customs and Border Protection (CBP) and contains a wealth of information about apprehensions and contraband seizures in border sectors.

We will be using the Sector Profile 2017 PDF which has information in four tables, three of which we'll scrape and then combine together. The data was downloaded from the US Customs and Border Protection "Stats and Summaries" page here.[2] If you're interested in using more of their data, some of it has been cleaned and made available here.[3]

The file we want to use is called "usbp_stats_fy2017_sector_profile.pdf" and has four tables in the PDF. Let's take a look at them one at a time, understanding what variables are available, and what units each row is in. Then we'll start scraping the tables.

The first table is "Sector Profile - Fiscal Year 2017 (Oct. 1st through Sept. 30th)". Before we even look down more at the table, the title is important. It is for fiscal year 2017, not calendar year 2017, which is more common in the data we usually use. This is important if we ever want to merge this data with other data sets. If possible, we would have to get data that is monthly so we can just use October 2016 through September 2017 to match up properly.

---

[1] https://github.com/jacobkap/crimebythenumbers/tree/master/data
[2] https://www.cbp.gov/newsroom/media-resources/stats
[3] https://www.openicpsr.org/openicpsr/project/109522/version/V2/view

United States Border Patrol

Sector Profile - Fiscal Year 2017 (Oct. 1st through Sept. 30th)

SECTOR	Agent Staffing"	Apprehensions	Other Than Mexican Apprehensions	Marijuana (pounds)	Cocaine (pounds)	Accepted Prosecutions	Assaults	Rescues	Deaths
Miami	111	2,280	1,646	2,253	231	292	1	N/A	N/A
New Orleans	63	920	528	21	6	10	0	N/A	N/A
Ramey	38	388	387	3	2,932	89	0	N/A	N/A
Coastal Border Sectors Total	212	3,588	2,561	2,277	3,169	391	1	N/A ****	N/A ****
Blaine	296	288	237	0	0	9	0	N/A	N/A
Buffalo	277	447	293	228	2	37	2	N/A	N/A
Detroit	408	1,070	322	124	0	85	1	N/A	N/A
Grand Forks	189	496	202	0	0	40	2	N/A	N/A
Havre	183	39	28	98	0	2	0	N/A	N/A
Houlton	173	30	30	17	0	2	0	N/A	N/A
Spokane	230	208	67	68	0	24	0	N/A	N/A
Swanton	292	449	359	531	1	103	6	N/A	N/A
Northern Border Sectors Total	2,048	3,027	1,538	1,086	3	302	11	N/A ****	N/A ****
Big Bend (formerly Marfa)	500	6,002	3,346	40,852	45	2,847	11	26	1
Del Rio	1,391	13,476	6,156	9,482	62	8,022	12	99	18
El Centro	870	18,633	5,812	5,554	484	1,413	34	4	2
El Paso	2,182	25,193	15,337	34,189	140	6,996	54	44	8
Laredo	1,666	25,460	7,891	69,535	757	6,119	31	1,054	83
Rio Grande Valley (formerly McAllen)	3,130	137,562	107,909	260,020	1,192	7,979	422	1,190	104
San Diego	2,199	26,086	7,060	10,985	2,903	3,099	84	48	4
Tucson	3,691	38,657	12,328	397,090	331	20,963	93	750	72
Yuma	859	12,847	10,139	30,181	261	2,367	33	6	2
Southwest Border Sectors Total"	16,605	303,916	175,978	857,888	6,174	59,805	774	3,221	294
Nationwide Total"*	19,437	310,531	180,077	861,231	9,346	60,498	786	3,221	294

* *Agent staffing statistics depict FY17 on-board personnel data as of 9/30/2017*

** *Southwest Border Sectors staffing statistics include: Big Bend, Del Rio, El Centro, El Paso, Laredo, Rio Grande Valley, San Diego, Tucson, Yuma, and the Special Operations Group*

*** *Nationwide staffing statistics include: All on-board Border Patrol agents in CBP*

**** *Rescue and Death statistics are not tracked for Northern and Coastal Border Sectors.*

Now if we look more at the table, we can see that each row is a section of the US border. There are three main sections - Coastal, Northern, and Southwest, with subsections of each also included. The bottom row is the sum of all these sections and gives us nationwide data. Many government data sets will be like this form with sections and subsections in the same table. Watch out when doing mathematical operations! Just summing any of these columns will give you triple the true value due to the presence of nationwide, sectional, and subsectional data.

There are 9 columns in the data other than the border section identifier. We have total apprehensions, apprehensions for people who are not Mexican citizens, marijuana and cocaine seizures (in pounds), the number of accepted prosecutions (presumably of those apprehended), and the number of CBP agents assaulted. The last two columns have the number of people rescued by CBP and the number of people who died (it is unclear from this data alone if this is solely people in custody or deaths during crossing the border). These two columns are also special as they only have data for the Southwest border.

The second table has a similar format with each row being a section or subsection. The columns now have the number of juveniles apprehended, subdivided by if they were accompanied by an adult or not, and the number of adults apprehended. The last column is total apprehensions which is also in the first table.

United States Border Patrol

Juvenile (0-17 Years Old) and Adult Apprehensions - Fiscal Year 2017 (Oct. 1st through Sept. 30th)

SECTOR	Accompanied Juveniles	Unaccompanied Juveniles	Total Juveniles	Total Adults	Total Apprehensions
Miami	19	42	61	2,219	2,280
New Orleans	1	22	23	897	920
Ramey	7	1	8	380	388
Coastal Border Sectors Total	27	65	92	3,496	3,588
Blaine	29	7	36	252	288
Buffalo	3	3	6	441	447
Detroit	5	11	16	1,054	1,070
Grand Forks	5	4	9	487	496
Havre	1	3	4	35	39
Houlton	1	8	9	21	30
Spokane	3	0	3	205	208
Swanton	18	10	28	421	449
Northern Border Sectors Total	65	46	111	2,916	3,027
Big Bend (formerly Marfa)	506	811	1,317	4,685	6,002
Del Rio	1,348	1,349	2,697	10,779	13,476
El Centro	968	1,531	2,499	16,134	18,633
El Paso	4,642	3,926	8,568	16,625	25,193
Laredo	477	2,033	2,510	22,950	25,460
Rio Grande Valley (formerly McAllen)	27,222	23,708	50,930	86,632	137,562
San Diego	1,639	1,551	3,190	22,896	26,086
Tucson	1,088	3,659	4,747	33,910	38,657
Yuma	3,241	2,867	6,108	6,739	12,847
Southwest Border Sectors Total	41,131	41,435	82,566	221,350	303,916
Nationwide Total	41,223	41,546	82,769	227,762	310,531

The third table follows the same format, and the new columns are number of apprehensions by gender.

United States Border Patrol

Apprehensions by Gender - Fiscal Year 2017 (Oct. 1st through Sept. 30th)

SECTOR	Female	Male	Total Apprehensions
Miami	219	2,061	2,280
New Orleans	92	828	920
Ramey	65	323	388
Coastal Border Sectors Total	376	3,212	3,588
Blaine	97	191	288
Buffalo	69	378	447
Detroit	78	992	1,070
Grand Forks	56	440	496
Havre	13	26	39
Houlton	17	13	30
Spokane	17	191	208
Swanton	106	343	449
Northern Border Sectors Total	453	2,574	3,027
Big Bend (formerly Marfa)	985	5,017	6,002
Del Rio	2,622	10,854	13,476
El Centro	2,791	15,842	18,633
El Paso	7,364	17,829	25,193
Laredo	3,651	21,809	25,460
Rio Grande Valley (formerly McAllen)	50,306	87,256	137,562
San Diego	4,117	21,969	26,086
Tucson	4,693	33,964	38,657
Yuma	4,328	8,519	12,847
Southwest Border Sectors Total	80,857	223,059	303,916
Nationwide Total	81,686	228,845	310,531

Finally, the fourth table is a bit different in its format. The rows are now variables, and the columns are the locations. In this table it doesn't include subsections, only border sections and the nationwide total. The data it has available are partially a repeat of the first table but with more drug types and the addition of the number of drug seizures and some firearm seizure information. As this table is formatted differently from the others, we won't scrape it in this lesson - but you can use the skills you'll learn to do so yourself.

United States Border Patrol

Apprehensions / Seizure Statistics - Fiscal Year 2017 (Oct. 1st through Sept. 30th)

Apprehension/Seizure Type	Coastal Border Sectors	Northern Border Sectors	Southwest Border Sectors	Nationwide Total
Apprehensions	3,588	3,027	303,916	310,531
Other Than Mexican Apprehensions	2,561	1,538	175,978	180,077
Marijuana (pounds)	2,277	1,066	857,888	861,231
Cocaine (pounds)	3,169	3	6,174	9,346
Heroin (ounces)	0	62	15,182	15,244
Methamphetamine (pounds)	23	32	10,273	10,328
Ecstasy (pounds)	0	0	1	1
Other Drugs* (pounds)	0	14	554	568
Marijuana Seizures	113	255	9,371	9,739
Cocaine Seizures	33	46	463	542
Heroin Seizures	0	29	219	248
Methamphetamine Seizures	2	68	809	879
Ecstasy Seizures	1	2	48	51
Other Drugs* Seizures	6	99	735	840
Conveyances	86	79	7,388	7,553
Firearms	9	45	369	423
Ammunition (rounds)	217	384	13,938	14,539
Currency (value)	$325,129	$374,282	$5,169,593	$5,869,004

*Other Drugs include: All USBP drug seizures excluding marijuana, cocaine, heroin, methamphetamine, and ecstasy (MDMA).
Coastal Border Sectors include: Miami, New Orleans, and Ramey, Puerto Rico.
Northern Border Sectors include: Blaine, Buffalo, Detroit, Grand Forks, Havre, Houlton, Spokane and Swanton.
Southwest Border Sectors include: Big Bend, Del Rio, El Centro, El Paso, Laredo, Rio Grande Valley, San Diego, Tucson, and Yuma.
Drug quantities are rounded to the nearest whole number

22.1   Scraping the first table

We've now seen all three of the tables that we want to scrape so we can begin the process of actually scraping them. Note that each table is very similar, meaning that we can reuse some code to scrape as well as to clean the data. That means that we will want to write some functions to make our work easier and avoid copy and pasting code.

We will start by using the pdf_text() function from the pdftools package to read the PDFs into R.

```
install.packages("pdftools")
```

```
library(pdftools)
# Using poppler version 22.04.0
```

We can assign the output of the `pdf_text()` function to the object *border_patrol*, and we'll use it for each table. The input to `pdf_text()` is the name of the PDF we want to scrape.

```
border_patrol <- pdf_text("data/usbp_stats_fy2017_sector_profile.pdf")
```

We can take a look at the `head()` of the result using `head(border_patrol)`.

```
head(border_patrol)
```

```
head(border_patrol)
#> [1] "                                    United States Border Patrol\n
#> [2] "                    United States Border Patrol\n
#> [3] "            United States Border Patrol\n
#> [4] "                United States Border Patrol\n
```

If you look closely in this huge amount of text output, you can see that it is a vector with each table being an element in the vector. We can see this further by checking the `length()` of "border_patrol", which tells us how many elements are in a vector.

```
length(border_patrol)
# [1] 4
```

It is four elements long, one for each table.

Looking at just the first element in *border_patrol* gives us all the values in the first table plus a few sentences at the end detailing some features of the table. At the end of each line (where in the PDF it should end but doesn't in our

data yet) there is a \n indicating that there should be a new line. We want to
use strsplit() to split at the \n.

```
border_patrol[1]
```

```
border_patrol[1]
#> [1] "                                            United States Border Patrol\n
```

The strsplit() function breaks up a string into pieces based on a value inside
of the string. Let's use the word "criminology" as an example. If we want to
split it by the letter "n" we'd have two results, "crimi" and "ology" as these
are the pieces of the word after breaking up "criminology" at letter "n".

```
strsplit("criminology", split = "n")
# [[1]]
# [1] "crimi" "ology"
```

Note that it deletes whatever value is used to break up the string.

Let's assign a new object with the value in the first element of *border_patrol*,
calling it *sector_profile* as that's the name of that table, and then using str-
split() on it to split it every \n. In effect this makes each line of the table
an element in a vector that we'll create rather than having the entire table
be a single long string as it is now. strsplit() returns a list so we will also
want to keep just the first element of that list using double square bracket
[[]] notation.

```
sector_profile <- border_patrol[1]
sector_profile <- strsplit(sector_profile, "\n")
sector_profile <- sector_profile[[1]]
```

Now we can look at the first six rows of this data.

```
head(sector_profile)
```

Notice that there is a lot of empty white space at the beginning of the rows. We want to get rid of that to make our next steps easier. We can use `trimws()` and put the entire *sector_profile* data in the (), and it'll remove any white space that is at the beginning or end of the string.

```
sector_profile <- trimws(sector_profile)
```

We have more rows than we want so let's look at the entire data and try to figure out how to keep just the necessary rows.

```
sector_profile
```

```
sector_profile
#>  [1] "United States Border Patrol"
#>  [2] "Sector Profile - Fiscal Year 2017 (Oct. 1st through Sept. 30th)"
#>  [3] ""
#>  [4] "Agent                          Other Than Mexican        Marijuana
#>  [5] "SECTOR                    Staffing*"
#>  [6] "Apprehensions"
#>  [7] "Apprehensions            (pounds)        (pounds)    Prosecutions"
#>  [8] "Assaults Rescues          Deaths"
#>  [9] ""
#> [10] "Miami                                    111            2,280
#> [11] "New Orleans                               63              920
#> [12] "Ramey                                     38              388
#> [13] "Coastal Border Sectors Total             212            3,588
#> [14] ""
#> [15] "Blaine                                   296              288
#> [16] "Buffalo                                  277              447
#> [17] "Detroit                                  408            1,070
#> [18] "Grand Forks                              189              496
#> [19] "Havre                                    183               39
#> [20] "Houlton                                  173               30
#> [21] "Spokane                                  230              208
#> [22] "Swanton                                  292              449
#> [23] "Northern Border Sectors Total          2,048            3,027
#> [24] "Big Bend (formerly Marfa)                500            6,002
#> [25] "Del Rio                                1,391           13,476
#> [26] "El Centro                                870           18,633
#> [27] "El Paso                                2,182           25,193
#> [28] "Laredo                                 1,666           25,460
#> [29] "Rio Grande Valley (formerly McAllen)   3,130          137,562
#> [30] "San Diego                              2,199           26,086
#> [31] "Tucson                                 3,691           38,657
#> [32] "Yuma                                     859           12,847
#> [33] "Southwest Border Sectors Total**      16,605          303,916
#> [34] "Nationwide Total***                   19,437          310,531
#> [35] "* Agent staffing statistics depict FY17 on-board personnel data as of 9/30/2017"
#> [36] "** Southwest Border Sectors staffing statistics include: Big Bend, Del Rio, El Centr
#> [37] "*** Nationwide staffing statistics include: All on-board Border Patrol agents in CBP
#> [38] "**** Rescue and Death statistics are not tracked for Northern and Coastal Border Sec
```

Based on the PDF, we want every row from Miami to Nationwide Total. But here we have several rows with the title of the table and the column names, and at the end we have the sentences with some details that we don't need.

To keep only the rows that we want, we can combine grep() and subsetting to find the rows from Miami to Nationwide Total and keep only those rows. We will use grep() to find which row has the text "Miami" and which has the text "Nationwide Total" and keep all rows between them (including those matched rows as well). Since each only appears once in the table we don't need to worry about handling duplicate results.

```
grep("Miami", sector_profile)
# [1] 10
```

```
grep("Nationwide Total", sector_profile)
# [1] 35
```

We'll use square bracket notation to keep all rows between those two values (including each value). Since the data is a vector, not a data.frame, we don't need a comma.

```
sector_profile <- sector_profile[grep("Miami", sector_profile):
grep("Nationwide Total", sector_profile)]
```

Note that we're getting rid of the rows that had the column names. It's easier to make the names ourselves than to deal with that mess. The data now has only the rows we want but still doesn't have any columns, it's currently just a vector of strings. We want to make it into a data.frame to be able to work on it like we usually do.

```
head(sector_profile)
```

```
head(sector_profile)
#> [1] "Miami                          111        2,280
#> [2] "New Orleans                     63         920
#> [3] "Ramey                           38         388
#> [4] "Coastal Border Sectors Total   212       3,588
#> [5] ""
#> [6] "Blaine                         296         288
```

When looking at this data it is clear that where the division between columns is supposed to be is a bunch of white space in each string. Take the first row for example, it says "Miami" then after lots of white spaces "111" than again with "2,280" and so on for the rest of the row. We'll use this pattern of columns differentiated by white space to make *sector_profile* into a data.frame.

We will use the function `str_split_fixed()` from the `stringr` package. This function is very similar to `strsplit()` except you can tell it how many columns to expect.

```
install.packages("stringr")
```

```
library(stringr)
```

The syntax of `str_split_fixed()` is similar to `strsplit()` except the new parameter of the number of splits to expect. The "_fixed" part of `str_split_fixed()` is that it expects the same number of splits (which in our case become columns) for every element in the vector that we input. Looking at the PDF shows us that there are 10 columns so that's the number we'll use. Our split will be " {2,}". That is, a space that occurs two or more times. Since there are sectors with spaces in their name, we can't have only one space, we need at least two. If you look carefully at the rows with sectors"Coastal Border Sectors Total" and "Northern Border Sectors Total", the final two columns actually do not have two spaces between them because of the amount of asterisks they have. Normally we'd want to fix this using `gsub()`, but those values will turn to NA anyway so we won't bother in this case.

```
sector_profile <- str_split_fixed(sector_profile, " {2,}", 10)
```

If we check the `head()` we can see that we have the proper columns now, but this still isn't a data.frame and has no column names.

```
head(sector_profile)
#        [,1]                                   [,2]  [,3]    [,4]
# [1,] "Miami"                                "111" "2,280" "1,646"
# [2,] "New Orleans"                          "63"  "920"   "528"
# [3,] "Ramey"                                "38"  "388"   "387"
# [4,] "Coastal Border Sectors Total" "212" "3,588" "2,561"
# [5,] ""                                     ""    ""      ""
# [6,] "Blaine"                               "296" "288"   "237"
```

```
#          [,5]     [,6]      [,7]   [,8] [,9]         [,10]
# [1,]  "2,253"  "231"     "292"  "1"  "N/A"        "N/A"
# [2,]  "21"     "6"       "10"   "0"  "N/A"        "N/A"
# [3,]  "3"      "2,932"   "89"   "0"  "N/A"        "N/A"
# [4,]  "2,277"  "3,169"   "391"  "1"  "N/A ****"   "N/A ****"
# [5,]  ""       ""        ""     ""   ""           ""
# [6,]  "0"      "0"       "9"    "0"  "N/A"        "N/A"
```

We can make it a data.frame just by putting it in `data.frame()`. And we can
assign the columns names using a vector of strings we can make. We'll use the
same column names as in the PDF but in lowercase and replacing spaces and
parentheses with underscores.

```
sector_profile <- data.frame(sector_profile)
names(sector_profile) <- c(
  "sector",
  "agent_staffing",
  "apprehensions",
  "other_than_mexican_apprehensions",
  "marijuana_pounds",
  "cocaine_pounds",
  "accepted_prosecutions",
  "assaults",
  "rescues",
  "deaths"
)
```

We have now taken a table from a PDF and successfully scraped it to a
data.frame in R. Now we can work on it as we would any other data set that
we've used previously.

```
head(sector_profile)
#                              sector agent_staffing apprehensions
# 1                             Miami            111         2,280
# 2                       New Orleans             63           920
# 3                             Ramey             38           388
# 4       Coastal Border Sectors Total            212         3,588
# 5
# 6                             Blaine            296           288
#    other_than_mexican_apprehensions marijuana_pounds
# 1                             1,646            2,253
```

```
# 2                          528              21
# 3                          387               3
# 4                        2,561           2,277
# 5
# 6                          237               0
#   cocaine_pounds accepted_prosecutions assaults   rescues
# 1             231                   292        1       N/A
# 2               6                    10        0       N/A
# 3           2,932                    89        0       N/A
# 4           3,169                   391        1 N/A ****
# 5
# 6               0                     9        0       N/A
#       deaths
# 1      N/A
# 2      N/A
# 3      N/A
# 4 N/A ****
# 5
# 6      N/A
```

To really be able to use this data we'll want to clean the columns to turn the values to numeric type, but we can leave that until later. For now let's write a function that replicates much of this work for the next tables.

22.2   Making a function

As we've done before, we want to take the code we wrote for the specific case of the first table in this PDF and turn it into a function for the general case of other tables in the PDF. Let's copy the code we used previously before we convert it to a function.

```
sector_profile <- border_patrol[1]
sector_profile <- trimws(sector_profile)
sector_profile <- strsplit(sector_profile, "\r\n")
sector_profile <- sector_profile[[1]]
sector_profile <- sector_profile[grep(
  "Miami",
  sector_profile
):
```

```
grep(
  "Nationwide Total",
  sector_profile
)]
sector_profile <- str_split_fixed(sector_profile, " {2,}", 10)
sector_profile <- data.frame(sector_profile)
names(sector_profile) <- c(
  "sector",
  "agent_staffing",
  "total_apprehensions",
  "other_than_mexican_apprehensions",
  "marijuana_pounds",
  "cocaine_pounds",
  "accepted_prosecutions",
  "assaults",
  "rescues",
  "deaths"
)
```

Since each table is so similar our function will only need a few changes in the above code to work for all three tables. The object *border_patrol* has all four of the tables in the data, so we need to say which of these tables we want - we can call the parameter `table_number`. Then each table has a different number of columns so we need to change the `str_split_fixed()` function to take a variable with the number of columns we input, a value we'll call `number_columns`. We rename each column to its proper name so we need to input a vector - which we'll call `column_names` - with the names for each column. Finally, we want to have a parameter where we enter in the data, which holds all of the tables, our object *border_patrol*, we can call this `list_of_tables` as it is fairly descriptive.

We do this as it is bad form (and potentially dangerous) to have a function that relies on an object that isn't explicitly put in the function. It we change our *border_patrol* object (such as by scraping a different file but calling that object *border_patrol*) and the function doesn't have that as an input, it will work differently than we expect. Since we called the object we scraped *sector_profile* for the first table, let's change that to *data* as not all tables are called Sector Profile.

```
scrape_pdf <- function(list_of_tables,
                       table_number,
                       number_columns,
                       column_names) {
  data <- list_of_tables[table_number]
```

```
    data <- trimws(data)
    data <- strsplit(data, "\n")
    data <- data[[1]]
    data <- data[grep("Miami", data):
    grep("Nationwide Total", data)]
    data <- str_split_fixed(data, " {2,}", number_columns)
    data <- data.frame(data)
    names(data) <- column_names

    return(data)
}
```

Now let's run this function for each of the three tables we want to scrape, changing the function's parameters to work for each table. To see what parameter values you need to input, look at the PDF itself or the screenshots in this lesson.

```
table_1 <- scrape_pdf(
  list_of_tables = border_patrol,
  table_number = 1,
  number_columns = 10,
  column_names = c(
    "sector",
    "agent_staffing",
    "total_apprehensions",
    "other_than_mexican_apprehensions",
    "marijuana_pounds",
    "cocaine_pounds",
    "accepted_prosecutions",
    "assaults",
    "rescues",
    "deaths"
  )
)
table_2 <- scrape_pdf(
  list_of_tables = border_patrol,
  table_number = 2,
  number_columns = 6,
  column_names = c(
    "sector",
    "accompanied_juveniles",
    "unaccompanied_juveniles",
    "total_juveniles",
```

```
     "total_adults",
     "total_apprehensions"
   )
 )
table_3 <- scrape_pdf(
  list_of_tables = border_patrol,
  table_number = 3,
  number_columns = 4,
  column_names = c(
    "sector",
    "female",
    "male",
    "total_apprehensions"
  )
)
```

We can use the function left_join() from the dplyr package to combine the three tables into a single object. In the first table there are some asterisks after the final two row names in the Sector column. For our match to work properly we need to delete them, which we can do using gsub().

```
table_1$sector <- gsub("\\*", "", table_1$sector)
```

Now we can run left_join(). left_join() will automatically join based on shared column names in the two data sets we are joining. In our case this is "sector" and "total_apprehensions." All we need to input into left_join() is the name of the data sets we want to join together. left_join() can only combine two data sets at a time so we'll first join table_1 and table_2 and then join table_3 with the result of the first join, which we'll call "final_data."

```
library(dplyr)
#
# Attaching package: 'dplyr'
# The following objects are masked from 'package:stats':
#
#     filter, lag
# The following objects are masked from 'package:base':
#
#     intersect, setdiff, setequal, union
final_data <- left_join(table_1, table_2)
# Joining, by = c("sector", "total_apprehensions")
```

```
final_data <- left_join(final_data, table_3)
# Joining, by = c("sector", "total_apprehensions")
```

Let's take a look at the head() of this combined data.

```
head(final_data)
#                          sector agent_staffing
# 1                         Miami            111
# 2                   New Orleans             63
# 3                         Ramey             38
# 4 Coastal Border Sectors Total            212
# 5
# 6                        Blaine            296
#   total_apprehensions other_than_mexican_apprehensions
# 1               2,280                            1,646
# 2                 920                              528
# 3                 388                              387
# 4               3,588                            2,561
# 5
# 6                 288                              237
#   marijuana_pounds cocaine_pounds accepted_prosecutions
# 1            2,253            231                    292
# 2               21              6                     10
# 3                3          2,932                     89
# 4            2,277          3,169                    391
# 5
# 6                0              0                      9
#   assaults rescues    deaths accompanied_juveniles
# 1        1     N/A       N/A                    19
# 2        0     N/A       N/A                     1
# 3        0     N/A       N/A                     7
# 4        1 N/A **** N/A ****                    27
# 5                                            <NA>
# 6        0     N/A       N/A                    29
#   unaccompanied_juveniles total_juveniles total_adults
# 1                      42              61        2,219
# 2                      22              23          897
# 3                       1               8          380
# 4                      65              92        3,496
# 5                    <NA>            <NA>         <NA>
# 6                       7              36          252
#   female  male
# 1    219 2,061
```

```
# 2      92    828
# 3      65    323
# 4     376  3,212
# 5    <NA>   <NA>
# 6      97    191
```

In one data set we now have information from three separate tables in a PDF. We have now scraped three different tables from a PDF and turned them into a single data set, turning the PDF into actually usable (and useful) data!

23

More scraping tables from PDFs

For this chapter you'll need the following files, which are available for download here[1]: AbbreRptCurrent.pdf and PregnantFemaleReportingCurrent.pdf.

In Chapter 22 we used the package `pdftools` to scrape tables on arrests/seizures from the US Customs and Border Protection that were only available in a PDF. Given the importance of PDF scraping, in this chapter we'll continue working on scraping tables from PDFs. Here, we will use the package `tabulizer`, which has a number of features making it especially useful for grabbing tables from PDFs.

One issue, which we saw in Chapter 22, is that the table may not be the only thing on the page - the page could also have a title, page number, etc. When using `pdftools` we use regular expressions and subsetting to remove all the extra lines. Using `tabulizer` we can simply say (through a handy function) that we only want a part of the page, so we only scrape the table itself. For more info about the `tabulizer` package please see its site here.[2]

23.1 Texas jail data

For this chapter we'll scrape data from the Texas Commission on Jail Standards - Abbreviated Population Report. This is a report that shows monthly data on people incarcerated in jails for counties in Texas. This PDF is 9 pages long because of how many counties there are in Texas. Let's take a look at what the first page looks like. If you look at the PDF yourself you'll see that every page follows the format of the 1st page, which greatly simplifies our scrape. The data is in county-month units, which means that each row of data has info for a single county in a single month. We know that because the first column is "County," and each row is a single county (this is not true in every case. For example, on page 3 there are the rows "Fannin 1(P)" and "Fannin 2(P)", possibly indicating that there are two jails in that county. It is unclear from this PDF what the "(P)" means.). For knowing that the data

---

[1]https://github.com/jacobkap/crimebythenumbers/tree/master/data
[2]https://docs.ropensci.org/tabulizer/

is monthly, the title of this document says "for 06/01/2020" indicating that it is for that date, though this doesn't by itself mean the data is monthly - it could be daily based only on this data.

To know for sure that it is monthly data we'd have to go to the original source on the Texas Commission on Jail Standards website here.[3] On this page it says that "Monthly population reports are available for review below," which tells us that the data is monthly. It's important to know the unit so you can understand the data properly - primarily so you know what kinds of questions you can answer. If someone asks whether yearly trends on jail incarceration change in Texas, you can answer that with this data. If they ask whether more people are in jail on a Tuesday than on a Friday, you can't.

Just to understand what units our data is in we had to look at both the PDF itself and the site it came from. This kind of multi-step process is tedious but often necessary to truly understand your data. And even now we have questions - what does the (P) that's in some rows mean? For this we'd have to email or call the people who handle the data and ask directly. This is often the easiest way to answer your question, though different organizations have varying speeds in responding - if ever.

Now let's look at what columns are available. It looks like each column is the number of people incarcerated in the jail, broken down into categories of people. For example, the first two columns after County are "Pretrial Felons" and "Conv. Felons" so those are probably how many people are incarcerated who are awaiting trial for a felony and those already convicted of a felony. The other columns seem to follow this same format until the last few ones, which describe the jails' capacity (i.e. how many people they can hold), what percent of capacity they are at, and specifically how many open beds they have.

---

[3]https://www.tcjs.state.tx.us/historical-population-reports/#1580454195676-420dac a6-0a306

Texas Commission on Jail Standards - Abbreviated Population Report for 06/01/2020

County	Pretrial Felons	Conv. Felons	Conv. Felons Sentenced to County Jail time	Parole Violators	Parole Violators with a New Charge	Pretrial Misd.	Conv. Misd.	Bench Warrants	Federal	Pretrial SJF	Conv. SJF Sentenced to Co. Jail Time	Conv. SJF Sentenced to State Jail	Total Others	Total Local	Total Contract	Total Population	Total Capacity	% of Capacity	Available Beds
Anderson	81	13	3	1	5	12	1	0	0	21	0	1	0	138	0	138	300	46.00	132
Andrews	23	11	0	2	4	11	0	0	0	5	0	6	0	35	0	35	50	70.00	10
Angelina	79	35	4	6	0	14	0	3	0	23	0	3	1	168	0	168	279	60.22	63
Aransas	23	10	0	2	6	7	0	6	73	2	0	0	0	56	73	129	212	60.85	62
Archer	12	3	0	0	1	3	1	1	2	5	0	0	1	26	9	35	48	72.92	0
Armstrong	1	1	0	0	0	0	0	0	0	0	0	0	0	2	0	2	8	25.00	0
Atascosa	54	2	0	8	4	21	0	5	0	29	0	4	0	127	29	156	250	62.40	69
Austin	17	2	0	1	3	3	0	0	0	5	0	0	2	32	0	32	90	35.56	49
Bailey	10	2	0	0	0	4	0	0	56	0	0	2	1	19	59	78	96	81.25	0
Bandera	26	8	0	1	3	3	2	0	0	1	0	1	0	45	11	56	96	58.33	30
Bastrop	111	4	0	13	11	22	1	4	70	17	1	0	0	184	72	256	400	64.00	104
Baylor	7	1	0	0	0	2	0	0	0	0	0	0	0	0	0	0	0	0.00	0
Bee	24	5	1	0	8	4	0	2	21	0	0	0	2	46	21	67	128	52.34	48
Bell	365	74	1	32	53	55	13	1	11	66	0	10	3	672	11	683	1184	57.69	383
Bexar	1552	225	56	268	302	253	10	52	2	287	5	52	379	3461	11	3472	5108	67.97	1125
Blanco	7	2	0	1	0	1	0	0	0	1	0	0	0	12	0	12	56	21.43	38
Borden	1	0	0	0	0	2	0	0	0	0	0	0	0	0	0	0	0	0.00	0
Bosque	4	1	0	0	1	1	1	1	0	1	0	0	2	12	0	12	64	18.75	46
Bowie (P)	31	29	0	2	5	22	4	2	47	24	2	4	1	126	490	616	921	66.88	213
Brazoria	219	167	10	17	29	34	18	15	0	96	0	35	14	654	0	654	1170	55.90	399
Brazos	200	78	3	16	66	37	8	6	0	52	0	20	24	510	0	510	1089	46.83	470
Brewster	3	0	0	0	0	0	0	0	38	0	0	0	0	3	38	41	56	73.21	0
Briscoe	0	0	0	0	0	0	0	0	0	0	0	0	0	0	0	0	0	0.00	0
Brooks	15	1	1	1	0	0	0	0	0	0	0	0	0	18	0	18	36	50.00	14
Brooks (P)	0	0	0	0	0	0	0	0	164	0	0	0	0	0	408	408	652	62.58	179
Brown	70	20	0	7	20	9	0	2	0	0	3	0	2	133	7	140	196	71.43	30
Burleson	19	2	0	2	0	3	0	0	0	0	0	0	1	27	0	27	96	28.13	59
Burnet	57	23	1	5	9	3	0	1	0	10	1	0	0	110	158	268	595	45.04	268
Caldwell	89	4	0	3	2	28	1	2	19	13	0	3	0	143	21	164	301	54.49	107

6/9/2020   Page 1 of 9

Now that we've familiarized ourselves with the data, let's begin scraping this data using `tabulizer`. If you don't have this package installed, you'll need to install it using `install.packages("tabulizer")`. Then we'll need to run `library(tabulizer)`.

```
install.packages("tabulizer")
```

```
library(tabulizer)
```

The main function that we'll be using from the `tabulizer` package is `extract_tables()`. In the parentheses we need to put the name of our PDF (in quotes). This function basically looks at a PDF page, figures out which part of the page is a table, and then scrapes just that table. As we'll see, it's not always perfect at figuring out what part of the page is a table so we can also tell it exactly where to look. You can look at all of the features of `extract_tables()` by running `help(extract_tables)`.

```
data <- extract_tables(file = "data/AbbreRptCurrent.pdf")
```

Normally we'd now look at the head() of our data object, but if we did that it would print out a very large amount of information. Instead, we'll check how long our object is using length() which tells us how many elements a vector or list has. We'll also check what type of data it is since different types of data (e.g. vector, data.frame) operate differently.

```
length(data)
# [1] 18
is(data)
# [1] "list"    "vector"
```

We learn that it is a list of length of 18, or has 18 elements in it. Why is this? We have 9 pages so it is reasonable that we would have 9 lists since we have one table per page, but we shouldn't have 18 tables.

Back in Section 3.3.3 I said that lists are one of the data types that we don't have to worry about as we don't use them much in this book. That's still true. Our data object is a list, and we want to convert this to a data.frame as quickly as possible. The important thing to know when interacting with a list is that subsetting here uses two pairs of square brackets [[]] instead of one pair of square brackets for a normal vector.

Let's look again at just the first table in our object, subsetting using [[1]].

```
data[[1]]
#          [,1]       [,2]       [,3]       [,4]
# [1,] ""         ""         ""         "Conv. Felons"
# [2,] ""         ""         ""         "Sentenced to"
# [3,] ""         ""         ""         "County Jail"
# [4,] ""         "Pretrial" "Conv."    ""
# [5,] ""         ""         ""         "time"
# [6,] "County"   "Felons"   "Felons"   ""
#          [,5]       [,6]       [,7]       [,8]     [,9]
# [1,] ""         "Parole"      ""         ""       ""
# [2,] ""         "Violators"   ""         ""       ""
# [3,] ""         "with a New"  ""         ""       ""
# [4,] "Parole"   ""            "Pretrial" "Conv."  "Bench"
# [5,] ""         "Charge"      ""         ""       ""
# [6,] "Violators" ""           "Misd."    "Misd."  "Warrants"
#          [,10]      [,11]      [,12]      [,13]
```

```
# [1,] ""          ""          "Conv. SJF"  "Conv."
# [2,] ""          ""          "Sentenced"  "SJF"
# [3,] ""          ""          "to Co. Jail" "Sentenced"
# [4,] ""          "Pretrial" ""            ""
# [5,] ""          ""          "Time"       "to State Jail"
# [6,] "Federal"  "SJF"       ""            ""
#       [,14]     [,15]   [,16]    [,17]        [,18]
# [1,] ""          ""      ""       ""           ""
# [2,] ""          ""      ""       ""           ""
# [3,] ""          ""      ""       ""           ""
# [4,] "Total"    "Total" "Total"  "Total"      "Total"
# [5,] ""          ""      ""       ""           ""
# [6,] "Others"   "Local" "Contract" "Population" "Capacity"
#       [,19]      [,20]
# [1,] ""          ""
# [2,] ""          ""
# [3,] ""          ""
# [4,] "% of"     "Available"
# [5,] ""          ""
# [6,] "Capacity" "Beds"
```

The results from `data[[1]]` provide some answers. It has the right number of columns but only 6 rows! This is our first table so it should be the entire table we can see on page 1. Instead, it appears to be just the column names, with 6 rows because some column names are on multiple rows. Here's the issue, we can read the table and easily see that the column names may be on multiple rows but belong together, and that they are part of the table. `tabulizer` can't see this obvious fact. It must rely on a series of rules to indicate what is part of a table and what isn't.

For example, having white space between columns and thin black lines around rows tells it where each row and column is. Our issue is that the column names appear to just be text until there is a thick black line and (in `tabulizer`'s mind) the table begins, so it keeps the column name part separate from the rest of the table. Now let's look closer at the second element in our data object and see if it is correct for the table on page 1 of our PDF.

```
head(data[[2]])
#       [,1]        [,2] [,3] [,4] [,5] [,6] [,7] [,8] [,9]
# [1,] "Anderson"  "81" "13" "3"  "1"  "5"  "12" "1"  "0"
# [2,] "Andrews"   "23" "11" "0"  "2"  "4"  "11" "0"  "0"
# [3,] "Angelina"  "79" "35" "4"  "6"  "0"  "14" "0"  "3"
# [4,] "Aransas"   "23" "10" "0"  "2"  "6"  "7"  "0"  "6"
```

```
# [5,] "Archer"     "12" "3"  "0"  "0"  "1"  "3"  "1"  "1"
# [6,] "Armstrong" "1"  "1"  "0"  "0"  "0"  "0"  "0"  "0"
#         [,10] [,11] [,12] [,13] [,14] [,15] [,16] [,17] [,18]
# [1,] "0"   "21"  "0"   "1"   "0"   "138" "0"   "138" "300"
# [2,] "0"   "5"   "0"   "6"   "0"   "35"  "0"   "35"  "50"
# [3,] "0"   "23"  "0"   "3"   "1"   "168" "0"   "168" "279"
# [4,] "73"  "2"   "0"   "0"   "0"   "56"  "73"  "129" "212"
# [5,] "2"   "5"   "0"   "0"   "1"   "26"  "9"   "35"  "48"
# [6,] "0"   "0"   "0"   "0"   "0"   "2"   "0"   "2"   "8"
#         [,19]   [,20]
# [1,] "46.00" "132"
# [2,] "70.00" "10"
# [3,] "60.22" "83"
# [4,] "60.85" "62"
# [5,] "72.92" "0"
# [6,] "25.00" "0"
tail(data[[2]])
#         [,1]        [,2] [,3] [,4] [,5] [,6] [,7] [,8] [,9]
# [24,] "Brooks"     "15" "1"  "1"  "1"  "0"  "0"  "0"  "0"
# [25,] "Brooks (P)" "0"  "0"  "0"  "0"  "0"  "0"  "0"  "0"
# [26,] "Brown"      "70" "20" "0"  "7"  "20" "9"  "0"  "2"
# [27,] "Burleson"   "19" "2"  "0"  "2"  "0"  "3"  "0"  "0"
# [28,] "Burnet"     "57" "23" "1"  "5"  "9"  "3"  "0"  "1"
# [29,] "Caldwell"   "89" "4"  "0"  "3"  "2"  "26" "1"  "2"
#         [,10] [,11] [,12] [,13] [,14] [,15] [,16] [,17] [,18]
# [24,] "0"   "0"   "0"   "0"   "0"   "18"  "0"   "18"  "36"
# [25,] "164" "0"   "0"   "0"   "0"   "0"   "408" "408" "652"
# [26,] "0"   "0"   "3"   "0"   "2"   "133" "7"   "140" "196"
# [27,] "0"   "0"   "0"   "0"   "1"   "27"  "0"   "27"  "96"
# [28,] "0"   "10"  "1"   "0"   "0"   "110" "158" "268" "595"
# [29,] "19"  "13"  "0"   "3"   "0"   "143" "21"  "164" "301"
#         [,19]   [,20]
# [24,] "50.00" "14"
# [25,] "62.58" "179"
# [26,] "71.43" "36"
# [27,] "28.13" "59"
# [28,] "45.04" "268"
# [29,] "54.49" "107"
```

We're looking just at the head() and tail() to get the first and last six rows as otherwise we'd print out all 29 rows in that table. When you are exploring your own data, you'll probably want to be more thorough and ensure that rows around the middle are also correct - but this is a good first pass. If you look at the output we just printed out and compare it to the PDF, you'll see that the scrape was successful. Every row is where it should be and the columns are correct - unlike when using `pdftools()`, we have the results already in proper columns.

One thing to note is that this data isn't in a data.frame format, it's in a matrix. Matrices are the default output of `extract_tables()` though you can set it to output a data.frame by setting the parameter `output = "data.frame"`. In our case we actually wouldn't want that due to the issue of the column names. As shown below, outputting to a data.frame will automatically take the first row of data and convert that to column names. So now we have our first county as the column names, which is not correct. Note too that the function added "X" before the column names that are numbers. That's because column names cannot start with a number so the function tries to fix it by adding the "X" to the start.

```
data <- extract_tables(
    file = "data/AbbreRptCurrent.pdf",
    output = "data.frame"
)
head(data[[2]])
#      Anderson X81 X13 X3 X1 X5 X12 X1.1 X0 X0.1 X21 X0.2 X1.2
# 1     Andrews   23  11  0  2  4  11    0  0    0   5    0    6
# 2    Angelina   79  35  4  6  0  14    0  3    0  23    0    3
# 3     Aransas   23  10  0  2  6   7    0  6   73   2    0    0
# 4      Archer   12   3  0  0  1   3    1  1    2   5    0    0
# 5   Armstrong    1   1  0  0  0   0    0  0    0   0    0    0
# 6    Atascosa   54   2  0  8  4  21    0  5    0  29    0    4
#      X0.3 X138 X0.4 X138.1 X300 X46.00 X132
# 1       0   35    0     35   50  70.00   10
# 2       1  168    0    168  279  60.22   83
# 3       0   56   73    129  212  60.85   62
# 4       1   26    9     35   48  72.92    0
# 5       0    2    0      2    8  25.00    0
# 6       0  127   29    156  250  62.40   69
```

Let's rerun the `extract_tables()` function, this time keeping it as outputting a matrix.

```
data <- extract_tables(file = "data/AbbreRptCurrent.pdf")
```

Since the column names are the same on each page, we can set the names manually. There are 20 columns so this will be a lot of writing, but it's simpler and quicker than trying to do it programmatically. Since each table will have the same column names, we'll want to create a vector with the column names to use for every table. Following normal naming conventions, we'll make everything lowercase, and the only punctuation we'll use is an underscore.

```
column_names <- c(
  "county",
  "pretrial_felons",
  "conv_felons",
  "conv_felons_sentence_to_county_jail_time",
  "parole_violators",
  "parole_violators_with_a_new_charge",
  "pretrial_misd",
  "conv_misd",
  "bench_warrants",
  "federal",
  "pretrial_sjf",
  "conv_sjf_sentenced_to_co_jail_time",
  "conv_sjf_sentence_to_state_jail",
  "total_others",
  "total_local",
  "total_contract",
  "total_population",
  "total_capacity",
  "percent_of_capacity",
  "available_beds"
)
```

We can combine the results from this vector with that of the second table to have a complete table from page 1 of our PDF. We do this first by making the second element from our data object into a data.frame. Then we use names() and assign the column names to that of the vector of names we just made. Since this is the table from page 1 of the PDF, we'll call the object *page1_table*. We'll look just at the head() of our *page1_table* object.

```
page1_table <- data[[2]]
page1_table <- data.frame(page1_table)
names(page1_table) <- column_names
head(page1_table)
#      county pretrial_felons conv_felons
# 1  Anderson              81          13
# 2   Andrews              23          11
# 3  Angelina              79          35
# 4   Aransas              23          10
```

```
# 5    Archer               12          3
# 6 Armstrong                1          1
#     conv_felons_sentence_to_county_jail_time parole_violators
# 1                                          3                1
# 2                                          0                2
# 3                                          4                6
# 4                                          0                2
# 5                                          0                0
# 6                                          0                0
#    parole_violators_with_a_new_charge pretrial_misd
# 1                                    5            12
# 2                                    4            11
# 3                                    0            14
# 4                                    6             7
# 5                                    1             3
# 6                                    0             0
#    conv_misd bench_warrants federal pretrial_sjf
# 1         1              0       0           21
# 2         0              0       0            5
# 3         0              3       0           23
# 4         0              6      73            2
# 5         1              1       2            5
# 6         0              0       0            0
#    conv_sjf_sentenced_to_co_jail_time
# 1                                    0
# 2                                    0
# 3                                    0
# 4                                    0
# 5                                    0
# 6                                    0
#    conv_sjf_sentence_to_state_jail total_others total_local
# 1                                1            0         138
# 2                                6            0          35
# 3                                3            1         168
# 4                                0            0          56
# 5                                0            1          26
# 6                                0            0           2
#    total_contract total_population total_capacity
# 1               0             138            300
# 2               0              35             50
# 3               0             168            279
# 4              73             129            212
# 5               9              35             48
# 6               0               2              8
```

```
#    percent_of_capacity available_beds
# 1                46.00           132
# 2                70.00            10
# 3                60.22            83
# 4                60.85            62
# 5                72.92             0
# 6                25.00             0
```

Looking at the results, we've done this correctly. The values are right and the column names are correct. We've done it for one page but now must add the remaining pages. We'll do this through a for loop. We want to take the code we used above and loop through each of the tables we have. Since half of our tables are just the column names and not actual data, we need to skip those elements in our for loop. Luckily, our data follows a pattern where the first element is the column names from page 1, the second is the data from page 1, the third is the column names from page 2, the fourth is the data from page 2, and so on. So we need only every other value from 1 to 18, or every even number. We can get every other value using logical values, as shown in the next section, but since we only have 18 elements we'll just create the simple vector ourselves: `c(2, 4, 6, 8, 10, 12, 14, 16, 18)`.

For our for loop we can copy the code above but let's change the object name from *page1_table* to *temp* as each iteration will be of a different page so *page1_table* doesn't make sense.

```
for (i in c(2, 4, 6, 8, 10, 12, 14, 16, 18)) {
  temp <- data[[i]]
  temp <- data.frame(temp)
  names(temp) <- column_names
}
```

Running the above code runs our for loop successfully but doesn't assign the output anywhere. It just runs one iteration, assigns it to *temp*, and then overwrites *temp* for the next iteration. What we really want is a single object, which will end up having every single row of data from every page in one data.frame. To do this we make an empty data.frame by saying some object gets `data.frame()` without anything in the parentheses. And then for every iteration of the loop we add the data that is in temp to this empty data.frame (which will soon fill up with data).

By creating an empty data.frame at the start we avoid having to name any of the column names or say how many rows of data there will be. To add data to this data.frame each iteration we will use the function `bind_rows()` from `dplyr` which stacks data sets on top of each other. Let's first look at a simple

example of this before including it in our for loop. To use `bind_rows()` we put
two (or more) data.frames as the parameters, and it will return a single data
set with all rows stacked together. Let's create two data.frames that each have
the rows of `head(mtcars)` as a demonstration.

```
library(dplyr)
example1 <- head(mtcars)
example2 <- head(mtcars)
bind_rows(example1, example2)
#                        mpg cyl disp  hp drat    wt  qsec
# Mazda RX4...1         21.0   6  160 110 3.90 2.620 16.46
# Mazda RX4 Wag...2     21.0   6  160 110 3.90 2.875 17.02
# Datsun 710...3        22.8   4  108  93 3.85 2.320 18.61
# Hornet 4 Drive...4    21.4   6  258 110 3.08 3.215 19.44
# Hornet Sportabout...5 18.7   8  360 175 3.15 3.440 17.02
# Valiant...6           18.1   6  225 105 2.76 3.460 20.22
# Mazda RX4...7         21.0   6  160 110 3.90 2.620 16.46
# Mazda RX4 Wag...8     21.0   6  160 110 3.90 2.875 17.02
# Datsun 710...9        22.8   4  108  93 3.85 2.320 18.61
# Hornet 4 Drive...10   21.4   6  258 110 3.08 3.215 19.44
# Hornet Sportabout...11 18.7  8  360 175 3.15 3.440 17.02
# Valiant...12          18.1   6  225 105 2.76 3.460 20.22
#                       vs am gear carb
# Mazda RX4...1          0  1    4    4
# Mazda RX4 Wag...2      0  1    4    4
# Datsun 710...3         1  1    4    1
# Hornet 4 Drive...4     1  0    3    1
# Hornet Sportabout...5  0  0    3    2
# Valiant...6            1  0    3    1
# Mazda RX4...7          0  1    4    4
# Mazda RX4 Wag...8      0  1    4    4
# Datsun 710...9         1  1    4    1
# Hornet 4 Drive...10    1  0    3    1
# Hornet Sportabout...11 0  0    3    2
# Valiant...12           1  0    3    1
```

The data that is printed out has 12 rows, and in this example the first six
and the last six rows are identical. `bind_rows()` took the second object in the
parentheses (*example2*) and stacked it right below the last row in *example1*.
In this case the columns are already in the same order, but if they weren't,
`bind_rows()` is smart enough to arrange the columns in the second object to be
the same as the first object.

Now we can run our for loop and create a single data set with every row from our 9 pages of data. We start by creating our empty data.frame, and we'll call that *final*. At the end of our loop we say that *final* gets bind_rows(final, temp) meaning that temp is stacked to the bottom of *final* every time the loop runs. We'll end this code chunk by looking at head() and tail() of *final* to be sure it worked correctly.

```
final <- data.frame()
for (i in c(2, 4, 6, 8, 10, 12, 14, 16, 18)) {
  temp <- data[[i]]
  temp <- data.frame(temp)
  names(temp) <- column_names
  final <- bind_rows(final, temp)
}
# New names:
# * `` -> `...21`
head(final)
#        county pretrial_felons conv_felons
# 1  Anderson              81          13
# 2   Andrews              23          11
# 3  Angelina              79          35
# 4   Aransas              23          10
# 5    Archer              12           3
# 6 Armstrong               1           1
#   conv_felons_sentence_to_county_jail_time parole_violators
# 1                                        3                1
# 2                                        0                2
# 3                                        4                6
# 4                                        0                2
# 5                                        0                0
# 6                                        0                0
#   parole_violators_with_a_new_charge pretrial_misd
# 1                                  5            12
# 2                                  4            11
# 3                                  0            14
# 4                                  6             7
# 5                                  1             3
# 6                                  0             0
#   conv_misd bench_warrants federal pretrial_sjf
# 1         1              0       0           21
# 2         0              0       0            5
# 3         0              3       0           23
# 4         0              6      73            2
# 5         1              1       2            5
```

```
# 6              0               0      0            0
#   conv_sjf_sentenced_to_co_jail_time
# 1                               0
# 2                               0
# 3                               0
# 4                               0
# 5                               0
# 6                               0
#   conv_sjf_sentence_to_state_jail total_others total_local
# 1                               1            0          138
# 2                               6            0           35
# 3                               3            1          168
# 4                               0            0           56
# 5                               0            1           26
# 6                               0            0            2
#   total_contract total_population total_capacity
# 1              0             138            300
# 2              0              35             50
# 3              0             168            279
# 4             73             129            212
# 5              9              35             48
# 6              0               2              8
#   percent_of_capacity available_beds ...21
# 1             46.00            132 <NA>
# 2             70.00             10 <NA>
# 3             60.22             83 <NA>
# 4             60.85             62 <NA>
# 5             72.92              0 <NA>
# 6             25.00              0 <NA>
tail(final)
#          county pretrial_felons conv_felons
# 264      Yoakum                        6
# 265       Young                       19
# 266      Zapata                       15
# 267      Zavala                       16
# 268 Zavala (P)                        0
# 269               Total         29173
#     conv_felons_sentence_to_county_jail_time
# 264                                        1
# 265                                        5
# 266                                        1
# 267                                        0
# 268                                        0
# 269                                     5814
```

```
#       parole_violators parole_violators_with_a_new_charge
# 264                  0                                  0
# 265                  0                                  5
# 266                  0                                  0
# 267                  0                                  3
# 268                  0                                  0
# 269                383                               2700
#       pretrial_misd conv_misd bench_warrants federal
# 264               0         1              0       0
# 265               1         3              1       0
# 266               0         5              0       1
# 267               0         4              0       0
# 268               0         0              0       0
# 269            3180      3370            415     816
#       pretrial_sjf conv_sjf_sentenced_to_co_jail_time
# 264              0                                   2
# 265              0                                   6
# 266             58                                   5
# 267              0                                   0
# 268              0                                   0
# 269           4354                                4195
#       conv_sjf_sentence_to_state_jail total_others
# 264                                 0            0
# 265                                 1            0
# 266                                 0            0
# 267                                 0            0
# 268                                 0            0
# 269                               161         1186
#       total_local total_contract total_population
# 264            0              9               21
# 265            0             41                2
# 266            0             27               58
# 267            0             22               17
# 268            0              0                0
# 269         2790          53017             6696
#       total_capacity percent_of_capacity available_beds ...21
# 264              30                    48         62.50    13
# 265              43                   144         29.86    87
# 266              85                   240         35.42   131
# 267              39                    66         59.09    20
# 268               0                     0                   0
# 269           59713                 93991         63.53 24681
```

If you look closely at the final several rows you'll see that there is an extra column, and that the second column ("pretrial_felons") is blank for all of these rows. That's because when scraping the final page tabulizer incorrectly added an empty column between the first and second column, meaning that all columns to the right of the first column shifted once to the right. So all the values in "pretrial_felons" are actually in "conv_felons" and so on. The last column now is named "...21" since it is the 21st column, and that name was made automatically as our *column_names* object only has 20 values. This can occasionally happen, even if seemingly identical formatted pages like we have here. To fix something like this, we'd want to check every column and delete any that had all values be empty strings. I leave solving this to you. While it may be a challenge, at this point in the book you have the skills to do it.

23.2 Pregnant women incarcerated

We'll finish this chapter with another example of data from Texas - this time using data on the number of pregnant women booked in Texas county jails. This data has a unique challenge: it has 10 columns, but we want to make it have only 2. In the data (shown following), it starts with a column of county names, then a column of the number of pregnant women booked into that county's jail. Next is another column of county names - instead of continuing onto another page, this data just makes new columns when it runs out of room. We'll scrape this PDF using tabulizer() and then work to fix this multiple-column issue.

Pregnant Females Booked In Texas County Jails for 6/1/2020   Month Total: 305

County		County		County		County		County	
Anderson	0	Delta	0	Irion	0	Motley	0	Upton	0
Andrews	1	Denton	3	Jack	0	Nacogdoches	2	Uvalde	0
Angelina	0	DeWitt	0	Jackson	1	Navarro	2	Val Verde (P)	1
Aransas	0	Dickens	0	Jasper	0	Newton	0	Van Zandt	0
Archer	1	Dickens (P)	0	Jeff Davis	0	Newton (P)	0	Victoria	1
Armstrong	0	Dimmit	0	Jefferson	0	Nolan	2	Walker	1
Atascosa	0	Donley	0	Jefferson (P)	0	Nueces	4	Waller	0
Austin	0	Duval	0	Jim Hogg	0	Ochiltree	0	Ward	0
Bailey	0	Eastland	0	Jim Wells	0	Oldham	0	Washington	0
Bandera	0	Ector	3	Johnson	2	Orange	0	Webb	3
Bastrop	0	Edwards	0	Jones	0	Palo Pinto	1	Wharton	1
Baylor	0	El Paso	8	Karnes	0	Panola	0	Wheeler	0
Bee	0	Ellis	0	Karnes (P)	0	Parker	1	Wichita	3
Bell	9	Erath	0	Kaufman	3	Parmer	0	Wilbarger	0
Bexar	27	Falls	0	Kendall	1	Pecos	0	Willacy	0
Blanco	0	Fannin 1(P)	0	Kenedy	0	Polk	0	Williamson	2
Borden	0	Fannin 2(P)	2	Kent	0	Polk (P)	0	Wilson	0
Bosque	0	Fayette	0	Kerr	1	Potter	4	Winkler	0
Bowie (P)	0	Fisher	0	Kimble	0	Presidio	0	Wise	1
Brazoria	2	Floyd	0	King	0	Rains	0	Wood	0
Brazos	5	Foard	0	Kinney	0	Randall	0	Yoakum	0
Brewster	0	Fort Bend	4	Kleberg	1	Reagan	0	Young	0
Briscoe	0	Franklin	0	Knox	0	Real	0	Zapata	0
Brooks	0	Freestone	1	La Salle	0	Red River	0	Zavala	0
Brooks (P)	1	Frio (P)	0	Lamar	0	Reeves	0	Zavala (P)	0
Brown	1	Gaines	0	Lamb	0	Refugio	1		
Burleson	1	Galveston	5	Lampasas	0	Roberts	0		
Burnet 1(P)	5	Garza	0	Lavaca	0	Robertson	0		
Caldwell	0	Gillespie	0	Lee	0	Rockwall	0		
Calhoun	1	Glasscock	0	Leon	0	Runnels	0		
Callahan	0	Goliad	0	Liberty (P)	3	Rusk	0		
Cameron	6	Gonzales	1	Limestone	0	Sabine	0		
Camp	0	Gray	1	Lipscomb	0	San Augustine	0		
Carson	0	Grayson	4	Live Oak	4	San Jacinto	0		
Cass	0	Gregg	1	Llano	1	San Patricio	0		
Castro	0	Grimes	0	Loving	0	San Saba	0		
Chambers	0	Guadalupe	6	Lubbock	9	Schleicher	0		
Cherokee	0	Hale	1	Lynn	1	Scurry	0		
Childress	0	Hall	0	Madison	0	Shackelford	0		
Clay	0	Hamilton	0	Marion	0	Shelby	0		
Cochran	0	Hansford	0	Martin	0	Sherman	0		
Coke	0	Hardeman	0	Mason	0	Smith	9		
Coleman	0	Hardin	0	Matagorda	0	Somervell	0		
Collin	0	Harris	10	Maverick	1	Starr	0		

Notice that this data doesn't even have column names, so we'll have to make them ourselves. This is always a bit risky as maybe next month the table will change, and if we hard-code any column names, we'll either have code that breaks or - much more dangerous - mislabel the columns without noticing. In cases like this we have no other choice, but if you intend to scrape PDFs that regularly update (such as when a new month of data comes out) be careful about situations like this.

We'll start scraping this PDF using the standard `extract_tables()` function without any parameters other than the file name. This is usually a good start since it's quick and often works - and if it doesn't, we haven't lost much time checking. Since we know `extract_tables()` will return a list by default, we'll assign the result of `extract_tables()` to an object called `data` and then just pull the first element (the only element if this scrape works properly) from that list. And to see how the scraping went, we'll look at the `head()` of the data.

```
data <- extract_tables(file = "data/PregnantFemaleReportingCurrent.pdf")
data <- data[[1]]
head(data)
#        [,1]        [,2] [,3]       [,4] [,5]       [,6]
# [1,]  "Anderson"  "0"  "Delta"    "0"  "Irion"    "0"
# [2,]  "Andrews"   "1"  "Denton"   "3"  "Jack"     "0"
# [3,]  "Angelina"  "0"  "DeWitt"   "0"  "Jackson"  "1"
```

```
# [4,] "Aransas"    "0"  "Dickens"      "0"  "Jasper"     "0"
# [5,] "Archer"     "1"  "Dickens (P)"  "0"  "Jeff Davis" "0"
# [6,] "Armstrong"  "0"  "Dimmit"       "0"  "Jefferson"  "0"
#       [,7]           [,8] [,9]           [,10]
# [1,] "Motley"        "0"  "Upton"          "0"
# [2,] "Nacogdoches"   "2"  "Uvalde"         "0"
# [3,] "Navarro"       "2"  "Val Verde (P)"  "1"
# [4,] "Newton"        "0"  "Van Zandt"      "0"
# [5,] "Newton (P)"    "0"  "Victoria"       "1"
# [6,] "Nolan"         "2"  "Walker"         "1"
```

If we check the output from the above code to the PDF, we can see that
it worked. Every column in the PDF is in our output, and the values were
scraped correctly. This is great! Now we want to make two columns - "county"
and "pregnant_females_booked" (or whatever you'd like to call it) - from
these 10. As usual with R, there are a few ways we can do this. We'll just do
this in two different ways.

First, since there are only 10 columns, we can just do it manually. We can use
square bracket [] notation to grab specific columns using the column number
(since the data is a matrix and not a data.frame we can't use dollar sign
notation even if we wanted to). We can see from the PDF that the county
columns are columns 1, 3, 5, 7, and 9. So can use a vector of numbers to get
that c(1, 3, 5, 7, 9).

```
head(data[, c(1, 3, 5, 7, 9)])
#       [,1]         [,2]          [,3]          [,4]
# [1,] "Anderson"   "Delta"       "Irion"       "Motley"
# [2,] "Andrews"    "Denton"      "Jack"        "Nacogdoches"
# [3,] "Angelina"   "DeWitt"      "Jackson"     "Navarro"
# [4,] "Aransas"    "Dickens"     "Jasper"      "Newton"
# [5,] "Archer"     "Dickens (P)" "Jeff Davis"  "Newton (P)"
# [6,] "Armstrong"  "Dimmit"      "Jefferson"   "Nolan"
#       [,5]
# [1,] "Upton"
# [2,] "Uvalde"
# [3,] "Val Verde (P)"
# [4,] "Van Zandt"
# [5,] "Victoria"
# [6,] "Walker"
```

Now again for the "pregnant_females_booked" columns, which are the even
numbers.

```
head(data[, c(2, 4, 6, 8, 10)])
#        [,1] [,2] [,3] [,4] [,5]
# [1,] "0"  "0"  "0"  "0"  "0"
# [2,] "1"  "3"  "0"  "2"  "0"
# [3,] "0"  "0"  "1"  "2"  "1"
# [4,] "0"  "0"  "0"  "0"  "0"
# [5,] "1"  "0"  "0"  "0"  "1"
# [6,] "0"  "0"  "0"  "2"  "1"
```

These results look right so we can make a data.frame using the `data.frame()`
and having the input be from the above code - removing the `head()` function
since we want every row. Conveniently, `data.frame()` allows us to name the
columns we are making so we'll name the two columns "county" and "preg-
nant_females_booked". We'll assign the result to an object that we'll call *data*
and check out the `head()` and `tail()` of that data.frame.

```
data <- data.frame(
  county = c(data[, c(1, 3, 5, 7, 9)]),
  pregnant_females_booked = c(data[, c(2, 4, 6, 8, 10)])
)
head(data)
#         county pregnant_females_booked
# 1    Anderson                       0
# 2     Andrews                       1
# 3    Angelina                       0
# 4     Aransas                       0
# 5      Archer                       1
# 6   Armstrong                       0
tail(data)
#         county pregnant_females_booked
# 295
# 296
# 297
# 298
# 299
# 300
```

These results look good! We now have only two columns, and the first six rows
(from `head()`) look right. Why are the last six rows all empty? Look back at the
PDF. The final two columns are shorter than the others, so `extract_tables()`
interprets them as empty strings. We can subset those away using a conditional
statement to remove any row with an empty string in either column. Since

we know that if there's an empty string in one of the columns it will also be there in the other, we only need to run this once.

```
data <- data[data$county != "", ]
head(data)
#        county pregnant_females_booked
# 1  Anderson                         0
# 2   Andrews                         1
# 3  Angelina                         0
# 4   Aransas                         0
# 5    Archer                         1
# 6 Armstrong                         0
tail(data)
#            county pregnant_females_booked
# 260         Wood                         0
# 261       Yoakum                         0
# 262        Young                         0
# 263       Zapata                         0
# 264       Zavala                         0
# 265 Zavala (P)                          0
```

Now the results from `tail()` look right.

We can now use the second method which will use logical values to only keep odd or even columns (as the columns we want are conveniently all odd or all even columns). First, I'm rerunning the code to scrape the PDF since now our *data* data set is already cleaned from above.

```
data <- extract_tables(file = "data/PregnantFemaleReportingCurrent.pdf")
data <- data[[1]]
```

We'll use a toy example now with a vector of numbers from 1 to 10 `1:10` which we can call *x*.

```
x <- 1:10
x
#  [1]  1  2  3  4  5  6  7  8  9 10
```

Now say we want every value of x and want to use logical values (also called Booleans) to get it. We need a vector of 10 values since we'd need one for every element in *x*. Specifically, we'd be using square bracket `[]` notation to subset (in this case not really a true subset since we'd return all the original values) and write ten TRUEs in the square brackets [].

```
x[c(TRUE, TRUE, TRUE, TRUE, TRUE, TRUE, TRUE, TRUE, TRUE)]
# [1]  1  2  3  4  5  6  7  8  9 10
```

If you're reading the code carefully, you might have notices that I only wrote nine TRUE values. Since R was expecting 10 values, when I only gave it nine, it started again from the beginning and used the first value in place of the expected tenth value. If we only wrote one TRUE value, R would just repeat that all 10 times.

```
x[TRUE]
# [1]  1  2  3  4  5  6  7  8  9 10
```

What happens when the value isn't always TRUE? It'll recycle it the exact same way. Let's try using now a vector c(TRUE, FALSE).

```
x[c(TRUE, FALSE)]
# [1] 1 3 5 7 9
```

It returns only the odd numbers. That's because the first value in our vector is TRUE so it returns the first value of *x*, which is 1. The next value is FALSE so it does not return the second value of *x*, which is 2. R then "recycles" our vector and uses the first value in our vector (TRUE) to interpret how to subset the third value of *x*. Since it's TRUE, it returns 3. But now the value for 4 is FALSE so it doesn't return it. The process repeats again until the end of the subset. Since every other value is returned, it returns only the odd numbers.

We can use R's method of "recycling" a vector that is shorter than it expects to solve our pregnant females booked issue. Indeed we can use this exact c(TRUE, FALSE) vector to select only the odd columns. Reversing it to c(FALSE, TRUE) gives us only the even columns.

So we'll copy over the code that made the data.frame last time and change the c(data[, c(1, 3, 5, 7, 9)] to c(data[, c(TRUE, FALSE)]) and the c(data[, c(2, 4, 6, 8, 10)]) to c(data[, c(FALSE, TRUE)]). Since the issue of empty strings is still there, we'll reuse the data <- data[data$county != "",] we made above to fix it.

```
data <- data.frame(
  county = c(data[, c(TRUE, FALSE)]),
  pregnant_females_booked = c(data[, c(FALSE, TRUE)])
)
data <- data[data$county != "", ]
head(data)
#        county pregnant_females_booked
# 1  Anderson                         0
# 2   Andrews                         1
# 3  Angelina                         0
# 4   Aransas                         0
# 5    Archer                         1
# 6 Armstrong                         0
tail(data)
#          county pregnant_females_booked
# 260        Wood                        0
# 261      Yoakum                        0
# 262       Young                        0
# 263      Zapata                        0
# 264      Zavala                        0
# 265 Zavala (P)                        0
```

23.3   Making PDF-scraped data available to others

You've now seen two examples of scraping tables from PDFs using the tabu-lizer() package and a few more examples from the pdftools package in Chapter 22. These chapters should get you started on most PDF scraping, but every PDF is different so don't rely on the functions alone to do all of the work. You'll still likely have to spend some time cleaning up the data afterwards to make it usable.

Given the effort you'll spend in scraping a PDF - and the relative rarity of this skill in criminology - I recommend that you help others by making your data available to the public. There are several current websites that let you do this, but I recommend openICPSR[4]. openICPSR lets people submit data for free (under a certain limit, 3GB per submission as of mid-2020 though you can ask for a limit increase) and has a number of features to make it easier to store and document the data. This includes a section to describe your data in text form, fill out tags to help people search for the data, and answer (optional)

---

[4]https://www.icpsr.umich.edu/web/pages/NACJD/index.html

questions on how the data was collected and the geographic and temporal scope of the data.

If you decide to update the data, it'll keep a link to your older submission so you essentially have versions of the data. When you update the data, I recommend having a section on the submission description describing the changes in each version. As an example of what it looks like when submitting data to openICPSR, below are a few images showing the submission page for one of my submissions that has many versions (and corresponding version notes).

Jacob Kaplan's Concatenated Files: Uniform Crime Reporting (UCR) Program Data: Supplementary Homicide Reports, 1976-2018 [PUBLISHED] ☑ Edit Project Header

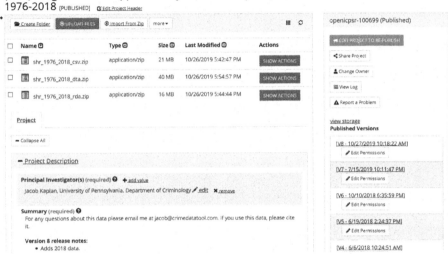

Summary (required) ❓

For any questions about this data please email me at jacob@crimedatatool.com. If you use this data, please cite it.

Version 8 release notes:
- Adds 2018 data.
- Changes source of data for years 1985-2018 to be directly from the FBI. 2018 data was received via email from the FBI, 2016-2017 is from the FBI who mailed me a DVD, and 1985-2015 data is from the FBI's Crime Data Explorer site (https://crime-data-explorer.fr.cloud.gov/downloads-and-docs).
- Adds .csv version of the data.
- Makes minor changes to value labels for consistency and to fix grammar.

Version 7 release notes:
- Changes project name to avoid confusing this data for the ones done by NACJD.

Version 6 release notes:
- Adds 2017 data.

Version 5 release notes:
- Adds 2016 data.
- Standardizes the "group" column which categorizes cities and counties by population.
- Arrange rows in descending order by year and ascending order by ORI.

Version 4 release notes:
- Fix bug where Philadelphia Police Department had incorrect FIPS county code.

Version 3 Release Notes:
- Merges data with LEAIC data to add FIPS codes, census codes, agency type variables, and ORI9 variable.
- Change column names for relationship variables from *offender_n_relation_to_victim_1* to *victim_1_relation_to_offender_n* to better indicate that all relationship are victim 1's relationship to each offender.
- Reorder columns.

This is a single file containing all data from the Supplementary Homicide Reports from 1976 to 2018. The Supplementary Homicide Report provides detailed information about the victim, offender, and circumstances of the murder. Details include victim and offender age, sex, race, ethnicity (Hispanic/not Hispanic), the weapon used, circumstances of the incident, and the number of both offenders and victims.

Years 1976-1984 were downloaded from NACJD, while more recent years are from the FBI. All files came as ASCII+SPSS Setup files and were cleaned using R. The "cleaning" just means that column names were

— Scope of Project

Subject Terms ❓
Do not copy/paste multiple terms into this field. Terms must be entered individually.
× SHR   × murder   × homicide   × supplementary homicide report

Geographic Coverage ❓ + add value
United States ✏ edit   ✕ remove

Time Period(s) ❓ + add value
1976 – 2018   ✏ edit   ✕ remove

Collection Date(s) ❓ + add value

Universe ❓
Victims of homicide in the United States between 1976 and 2018. ✏ edit   ✕ remove

Data Type(s) ❓
administrative records data   aggregate data   ✏ edit   ✕ remove

Collection Notes ❓
✏ edit

— Methodology

Response Rate ❓
✏ edit

Sampling ❓
✏ edit

Data Source ❓
United States Department of Justice. Federal Bureau of Investigation ✏ edit   ✕ remove

24

Geocoding

For this chapter you'll need the following file, which is available for download here[1]: san_francisco_active_marijuana_retailers.csv.

Several recent studies have looked at the effect of marijuana dispensaries on crime around the dispensary. For these analyses they find the coordinates of each crime in the city and see if it occurred in a certain distance from the dispensary. Many crime data sets provide the coordinates of where each crime occurred, however sometimes the coordinates are missing - and other data such as marijuana dispensary locations give only the address - meaning that we need a way to find the coordinates of these locations.

24.1   Geocoding a single address

In this chapter we will cover how to geocode addresses. Geocoding is the process of taking an address (e.g. 123 Main Street, Somewhere, CA, 12345) and getting the longitude and latitude coordinates of that address. With these coordinates we can then do spatial analyses on the data ranging from simply making a map and showing where each address is to merging these coordinates with some other spatial data (such as seeing which police district the address is in) and seeing how it relates to other variables, such as crime.

To do our geocoding, we're going to use the package `tidygeocoder` which greatly simplifies the work of geocoding addresses in R. For more information about this package, please see the package's site here.[2] If you've never used this package before you'll need to install it using `install.packages("tidygeocoder")`.

```
install.packages("tidygeocoder")
```

---

[1]https://github.com/jacobkap/crimebythenumbers/tree/master/data
[2]https://jessecambon.github.io/tidygeocoder/

Now we need to tell R that we want to use this package by running li-
brary(tidygeocoder).

```
library(tidygeocoder)
```

To geocode our addresses we'll use the helpfully named geocode() function
inside of tidygeocoder. For geocode() we input an address and it returns the
coordinates for that address. For our address we'll use "750 Race St. Philadelphia, PA 19106," which is the address of the Philadelphia Police Department
headquarters.

```
geocode("750 Race St. Philadelphia, PA 19106")
# Error: .tbl is not a dataframe. See ?geocode
```

As shown above, running geocode("750 Race St. Philadelphia, PA 19106") gives us
an error that tells us that ".tbl is not a dataframe." The issue is that geocode()
expects a data.frame (and .tbl is an abbreviation for tibble, which is a kind
of data.frame), but we entered only the string with our one address, not a
data.frame. For this function to work we need to enter two parameters into
geocode(): a data.frame (or something similar such as a tibble) and the name
of the column that has the addresses.[3] Since we need a data.frame, we'll make
one below. I'm calling it *address_to_geocode* and calling the column with
the address "address", but you can call both the data.frame and the column
whatever name you want.

```
address_to_geocode <- data.frame(
  address =
    "750 Race St. Philadelphia, PA 19106"
)
```

Now let's try again. We'll enter our data.frame *address_to_geocode* first and
then the name of our column which is "address".

```
geocode(address_to_geocode, address)
# # A tibble: 1 x 3
#   address                                lat   long
#   <chr>                                <dbl> <dbl>
# 1 750 Race St. Philadelphia, PA 19106  40.0 -75.2
```

---

[3]We can look at all of the parameters for this function by running the code help(geocode)
or ?geocode() to look at the functions Help page.

It worked, returning the same data.frame but with two additional columns with the latitude and longitude of that address.

You might be wondering why we put "address" into `geocode()` without quotes when usually when we talk about a column we need to do so in quotes. The simple answer is that the authors of the `tidygeocoder` package spent the time allowing users to input the column name either with or without quotes. Trying it again and now having "address" in quotes gives us the same result.

```
geocode(address_to_geocode, "address")
# # A tibble: 1 x 3
#   address                            lat  long
#   <chr>                            <dbl> <dbl>
# 1 750 Race St. Philadelphia, PA 19106  40.0 -75.2
```

There are two additional parameters that are important to talk about for this function, especially when you encounter an address that doesn't geocode properly.

First, there are actually multiple sources where you can enter an address and get the coordinates for that address. Just think about the big mapping apps or sites, such as Google Maps and Apple Maps. For these sources you can enter in the same address and you'll get different results. In most cases you'll get extremely similar coordinates, usually off only after a few decimals points, so they are functionally identical. But occasionally you'll have some addresses that can be geocoded through some sources but not others. This is because some sources have a more comprehensive list of addresses than others.

At the time of this writing the `tidygeocoder` package can handle geocoding from 13 different sources. For 10 of these, however, you need to setup an API key and some also require paying money (usually after a set number of addresses that it'll geocode for free each day). So here I'll just cover the three sources of geocoding that don't require any setup: "osm" (Open Street Map or OSM is similar to Google Maps), "census" (the US Census Bureau's geocoder), and "arcgis" (ArcGIS is a clunky mapping software that nonetheless has an excellent geocoder that R can use). To select which of these to use ("osm" is the default), you add the parameter "method" and set that equal to which one you want to use. As "osm" is the default we actually don't need to set it explicitly, but we'll do so anyways here as an example of the three geocoding sources we want to use.

```
example <- geocode(address_to_geocode, "address", method = "osm")
example
# # A tibble: 1 x 3
```

```
#   address                              lat  long
#   <chr>                                <dbl> <dbl>
# 1 750 Race St. Philadelphia, PA 19106  40.0 -75.2
```

```
example <- geocode(address_to_geocode, "address", method = "census")
example
# # A tibble: 1 x 3
#   address                              lat  long
#   <chr>                                <dbl> <dbl>
# 1 750 Race St. Philadelphia, PA 19106  40.0 -75.2
```

```
example <- geocode(address_to_geocode, "address", method = "arcgis")
example
# # A tibble: 1 x 3
#   address                              lat  long
#   <chr>                                <dbl> <dbl>
# 1 750 Race St. Philadelphia, PA 19106  40.0 -75.2
```

By default this function returns a tibble instead of a normal data.frame so it only shows one decimal point by default - though it doesn't actually round the number, merely shorten what it shows us. We can change the output back into a data.frame by using the `data.frame()` function. If you check each result after converting it to a data.frame you'll see that each set of coordinates are very slightly different, though for all purposes are the same location.

```
example <- geocode(address_to_geocode, "address", method = "arcgis")
example <- data.frame(example)
example
#                                       address      lat      long
# 1 750 Race St. Philadelphia, PA 19106 39.95488 -75.15205
```

Given how similar the coordinates are, you really only need to set the source of the geocoder in cases where one geocoder fails to find a match for the address.

The second important parameter is `full_results`, which is by default set to FALSE. When set to TRUE it gives more columns in the returning data.frame than just the longitude and latitude of that address. These columns differ for each geocoder source so we'll look at all three. I'll convert all of these results to a data.frame so it prints out all of the columns, and doesn't abbreviate results, which is how tibbles function. The `example$display_name <- NULL` isn't necessary, but I use it to remove a column that prints out an extremely long line for the location's full address and that looks bad in the print version of this book.

```
example <- geocode(address_to_geocode, "address",
  method = "osm", full_results = TRUE
)
example <- data.frame(example)
example$display_name <- NULL
example
#                                     address     lat      long
# 1 750 Race St. Philadelphia, PA 19106 39.95506 -75.15217
#     place_id
# 1 304083147
#                                                           licence
# 1 Data © OpenStreetMap contributors, ODbL 1.0. https://osm.org/copyright
#   osm_type     osm_id
# 1      way 626341043
#                                                       boundingbox
# 1 39.955010747193, 39.955110747193, -75.152217431613, -75.152117431613
#   class   type importance
# 1 place  house      -0.62
```

For OSM as a source we also get information about the address, such as what type of place it is, a bounding box which is a geographic area right around this coordinate, the address for those coordinates in the OSM database, and a bunch of other variables that don't seem very useful for our purposes such as the "importance" of the address. It's interesting that OSM classifies this address as a "house" as the headquarters of a major police department is quite a bit bigger than a house, so this is likely an misclassification of the type of address. The most important extra variable here is the address, called the "display_name".

Sometimes geocoders will be quite a bit off in their geocoding because they match the address you inputted incorrectly to one in their database. For example, if you input "123 Main Street" and the geocoder thinks you mean "123 Maine Street" you may be quite a bit off in the resulting coordinates. When you only get coordinates returned you won't know that the coordinates are wrong. Even if you know where an address is supposed to be it's hard to catch errors like this. If you're geocoding addresses in a single city and one point is in a different city (or completely different part of the world), then it's pretty clear that there's an error. But if the coordinates are simply in a wrong part of the city, but near other coordinates, then it's very hard to notice a problem.

So having an address to check against the one you inputted is a very useful way of validate the geocoding.

```
example <- geocode(address_to_geocode, "address",
  method = "census", full_results = TRUE
)
example <- data.frame(example)
example
#                                  address     lat     long
# 1 750 Race St. Philadelphia, PA 19106 39.95488 -75.1514
#                          matchedAddress tigerLine.side
# 1 750 RACE ST, PHILADELPHIA, PA, 19106                L
#   tigerLine.tigerLineId addressComponents.zip
# 1             131423677                 19106
#   addressComponents.streetName addressComponents.preType
# 1                         RACE
#   addressComponents.city addressComponents.preDirection
# 1           PHILADELPHIA
#   addressComponents.suffixDirection
# 1
#   addressComponents.fromAddress addressComponents.state
# 1                           700                      PA
#   addressComponents.suffixType addressComponents.toAddress
# 1                           ST                         798
#   addressComponents.suffixQualifier
# 1
#   addressComponents.preQualifier
# 1
```

The Census results are similar to the OSM results and also have the matched address to compare your inputted address to. Most of the columns are just the address broken into different pieces (street, city, state, etc.) so are mostly repeating the address again in multiple columns.

```
example <- geocode(address_to_geocode, "address",
  method = "arcgis", full_results = TRUE
)
example <- data.frame(example)
example
#                                  address     lat      long
# 1 750 Race St. Philadelphia, PA 19106 39.95488 -75.15205
#                             arcgis_address score
# 1 750 Race St, Philadelphia, Pennsylvania, 19106   100
```

```
#     location.x location.y extent.xmin extent.ymin extent.xmax
# 1   -75.15205   39.95488   -75.15305   39.95388   -75.15105
#     extent.ymax
# 1     39.95588
```

For the ArcGIS results we have the matched address again, and then an important variable called "score," which is basically a measure of how confident ArcGIS is that it matched the right address. Higher values are more confident, but in my experience anything under 90-95 confidence is an incorrect address. These results also repeat the longitude and latitude columns as "location.x" and "location.y" columns, and I'm not sure why they do so.

24.2   Geocoding San Francisco marijuana dispensary locations

So now that we can use the `geocoder()` function well, we can geocode every location in our marijuana dispensary data.

Let's read in the marijuana dispensary data, which is called "san_francisco_active_marijuana_retailers.csv" and call the object *marijuana*. Note the "data/" part in front of the name of the .csv file. This is to tell R that the file we want is in the "data" folder of our working directory. Doing this is essentially a shortcut to changing the working directory directly. For this book I keep all of the data files in a folder called "data" in my working directory. Unless you also have a folder called "data" in your working directory which has this file, please delete "data/" from the following code.

```
library(readr)
marijuana <- read_csv("data/san_francisco_active_marijuana_retailers.csv")
marijuana <- data.frame(marijuana)
```

Let's look at the top 6 rows.

```
head(marijuana)
#     License.Number              License.Type
# 1 C10-0000614-LIC Cannabis - Retailer License
# 2 C10-0000586-LIC Cannabis - Retailer License
# 3 C10-0000587-LIC Cannabis - Retailer License
```

```
# 4 C10-0000539-LIC Cannabis - Retailer License
# 5 C10-0000522-LIC Cannabis - Retailer License
# 6 C10-0000523-LIC Cannabis - Retailer License
#       Business.Owner        Business.Structure
# 1      Terry Muller Limited Liability Company
# 2     Jeremy Goodin                Corporation
# 3      Justin Jarin                Corporation
# 4 Ondyn Herschelle               Corporation
# 5       Ryan Hudson Limited Liability Company
# 6       Ryan Hudson Limited Liability Company
#                                       Premise.Address
# 1  2165 IRVING ST san francisco, CA 94122 County: SAN FRANCISCO
# 2 122 10TH ST SAN FRANCISCO, CA 941032605 County: SAN FRANCISCO
# 3   843 Howard ST SAN FRANCISCO, CA 94103 County: SAN FRANCISCO
# 4    70 SECOND ST SAN FRANCISCO, CA 94105 County: SAN FRANCISCO
# 5   527 Howard ST San Francisco, CA 94105 County: SAN FRANCISCO
# 6 2414 Lombard ST San Francisco, CA 94123 County: SAN FRANCISCO
#    Status Issue.Date Expiration.Date
# 1 Active  9/13/2019       9/12/2020
# 2 Active  8/26/2019       8/25/2020
# 3 Active  8/26/2019       8/25/2020
# 4 Active   8/5/2019        8/4/2020
# 5 Active  7/29/2019       7/28/2020
# 6 Active  7/29/2019       7/28/2020
#                   Activities Adult.Use.Medicinal
# 1 N/A for this license type            BOTH
# 2 N/A for this license type            BOTH
# 3 N/A for this license type            BOTH
# 4 N/A for this license type            BOTH
# 5 N/A for this license type            BOTH
# 6 N/A for this license type            BOTH
```

The column with the address is called *Premise Address*. Since the address county is always "County: SAN FRANCISCO" we can just gsub() out that entire string.

```
marijuana$Premise.Address <- gsub(
  " County: SAN FRANCISCO",
  "", marijuana$Premise.Address
)
```

Now let's make sure we did it right.

```
head(marijuana$Premise.Address)
# [1] "2165 IRVING ST san francisco, CA 94122"
# [2] "122 10TH ST SAN FRANCISCO, CA 941032605"
# [3] "843 Howard ST SAN FRANCISCO, CA 94103"
# [4] "70 SECOND ST SAN FRANCISCO, CA 94105"
# [5] "527 Howard ST San Francisco, CA 94105"
# [6] "2414 Lombard ST San Francisco, CA 94123"
```

To do the geocoding we'll just tell `geocode()` our data.frame name and the name of the column with the addresses. We'll assign the results back into the `marijuana` object. As noted earlier, we don't need to put the name of our column in quotes, but I like to do so because it is consistent with some other functions that require it. Running this code may take up to a minute because it's geocoding 33 different addresses.

```
marijuana <- geocode(marijuana, "Premise.Address")
```

Now it appears that we have longitude and latitude for every dispensary. We should check that they all look sensible.

```
summary(marijuana$long)
#    Min. 1st Qu. Median   Mean 3rd Qu.   Max.   NA's
#  -122.5  -122.4 -122.4 -122.4  -122.4 -122.4    10
```

```
summary(marijuana$lat)
#    Min. 1st Qu. Median   Mean 3rd Qu.   Max.   NA's
#   37.71   37.75  37.78  37.77   37.78  37.80    10
```

The minimum and maximum are very similar to each other for both longitude and latitude so that's a sign that it geocoded correctly. The 10 NA values mean that it didn't find a match for 10 of the addresses. Let's try again and now set `method` to "arcgis." which generally has a very high match rate. Before we do this let's just remove the entire latitude and longitude columns from our data. How the `geocode()` function works is that if we keep the "long" and "lat" columns that are currently in the data from when we just geocoded, when we run it again it'll make new columns that have nearly identical names. We usually want as few columns in our data as possible so there's no point having the "lat" column from the last geocode run with the 10 NAs and another "lat" (though slightly different, automatically chosen name) column from this time we run geocode().

We could also just geocode the 10 addresses that failed on the first run, but given that we'll only be geocoding a small number of addresses it won't take much extra time to have ArcGIS run it all. Running this function on just the NA rows requires a bit more work than just rerunning them all. In general, when the choice is between you spending time writing code and letting the computer do more work, let the computer do the work. And in general I'd recommend starting with ArcGIS as it is more reliable for geocoding. We'll remove the current coordinate columns by setting them each to NULL.

```
marijuana$long <- NULL
marijuana$lat <- NULL
marijuana <- geocode(marijuana, "Premise.Address",
  method = "arcgis"
)
```

And let's do the summary() check again.

```
summary(marijuana$long)
#    Min. 1st Qu.  Median    Mean 3rd Qu.    Max.
#  -122.5  -122.4  -122.4  -122.4  -122.4  -122.4
```

```
summary(marijuana$lat)
#    Min. 1st Qu.  Median    Mean 3rd Qu.    Max.
#   37.71   37.76   37.77   37.77   37.78   37.80
```

No more NAs, which means that we successfully geocoded our addresses. Another check is to make a simple scatterplot of the data. Since all of the data is from San Francisco, they should be relatively close to each other. If there are dots far from the rest, that is probably a geocoding issue.

```
plot(marijuana$long, marijuana$lat)
```

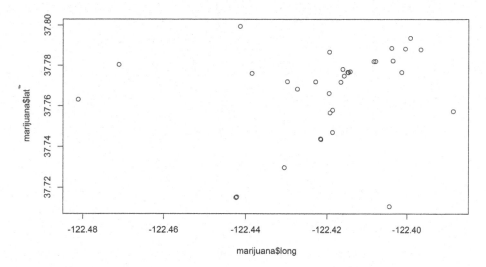

Most points are within a very narrow range so it appears that our geocoding worked properly.

Bibliography

Jain, H. C., Singh, P., and Agocs, C. (2000). Recruitment, selection and promotion of visible-minority and Aboriginal police officers in selected Canadian police services. *Canadian Public Administration*, 43(1):46–74.

Reaves, B. (1993). *Using NIBRS data to analyze violent crime.* US Department of Justice, Office of Justice Programs, Bureau of Justice.